GENETIC MECHANISMS OF SPECIATION
IN INSECTS

GENETIC MECHANISMS
OF SPECIATION IN INSECTS

SYMPOSIA HELD AT THE XIVTH INTERNATIONAL CONGRESS OF ENTOMOLOGY,
CANBERRA, AUSTRALIA AUGUST 22-30, 1972, SPONSORED BY
THE AUSTRALIAN ACADEMY OF SCIENCE AND THE
AUSTRALIAN ENTOMOLOGICAL SOCIETY

Edited by
M. J. D. WHITE

Springer-Science+Business Media, B.V.

Library of Congress Catalog Card Number 74-80531

ISBN 978-94-010-2250-7   ISBN 978-94-010-2248-4 (eBook)
DOI 10.1007/978-94-010-2248-4

Preface

Two Symposia on speciation in insects held at the Fourteenth International Congress of Entomology (Canberra, Australia, August 22-30, 1972) are included in this volume.

The first, on the more general topic of *Genetic Analysis of Speciation Mechanisms,* includes four papers on speciation in various groups of Diptera and Orthopteroid insects. The second symposium was devoted to the topic of *Evolution in the Hawaiian Drosophilidae*; it deals with the explosive speciation of a group of flies with specialized ecological requirements in the complex ecological habitats provided by a recent tropical volcanic archipelago. The Hawaiian Symposium, organized by Professor D. Elmo Hardy, is the latest outcome of a major collaborative research project involving over 20 scientists and about 125 technical assistants over a period of ten years.

Some recent books on evolution have taken the standpoint that the fundamental genetic mechanism of speciation is relatively uniform and stereotyped and, in particular, that the 'allopatric' model of its geographic component is universally valid. Certainly, this has been a rather generally accepted viewpoint on the part of students of vertebrate speciation. Workers on speciation in insects have tended, in general, to be less dogmatic and more willing to consider a variety of alternative models of speciation. Thus, in the present volume, several contributions adopt viewpoints which are unorthodox or novel. Only time will tell whether their conclusions will turn out to have been soundly based. In the meanwhile their authors have provided a wealth of morphological, ethological, ecological, biochemical and cytogenetic evidence which will have to be incorporated into the final synthesis. The *virilis* group of *Drosophila* in the northern hemisphere, the host-races of Tephritid flies, the morabine grasshoppers and eucalypt-phasmatids of Australia and the explosive speciation of the Hawaiian Drosophilidae will continue to be important for evolutionary theory for a long time to come. One may regret that in these entomological symposia there were no contributions from workers on Lepidoptera, Coleoptera and some other insect orders which have provided material for important evolutionary studies in the past few years. Mathematical and biometrical papers were also absent. One must hope that in future symposia of this kind other materials and additional lines of approach to the central problem will be abundantly represented. What the present volume does demonstrate clearly is that evolutionary studies on insects, both in the field and in the laboratory, utilizing a very wide variety of investigative techniques, are being very actively pursued at the present time.

M.J.D.W.

Melbourne Australia, March 17, 1973

Contents

I

GENETIC ANALYSIS OF SPECIATION MECHANISMS

The mechanism of sympatric host race formation in the true fruit flies (*Tephritidae*)

G. L. BUSH

Department of Zoology, University of Texas, Austin, Texas 78712

A lack of unanimity among evolutionary biologists concerning the processes involved in host race formation and speciation of parasitic insects of plants and animals has existed from the time Benjamin Walsh published his provocative paper on sympatric host race formation in phytophagous insects in 1864. Use of the term host race itself has created problems as it has been applied to taxons ranging from sibling species to biotypes. I will limit it to an infraspecific category generally applied to populations of a parasitic species which exhibit distinct genetically-based preferences for certain host plants.

The controversy in recent years has focused on three major unresolved questions. First, how fast can reproductive isolation evolve in obligate insect parasites? Second, is speciation always preceded by or dependent upon a major genetic revolution, and third, can speciation evolve sympatrically in the absence of any true geographic isolation as a result of ecological specialization on new host plants and animals or other novel ecological conditions? The viewpoint held by most evolutionary biologists is that host races develop the requisite reproductive isolating mechanisms only after long periods of physical isolation and considerable genetic alteration of the gene pool (See Mayr, 1963, for an excellent review). Some recent biologists disagree and contend or infer that in some groups only a few key genetic changes can lead to speciation in a matter of a few generations without physical barriers to gene flow (Smith, 1941; Thorpe, 1945; Haldane, 1959; Nowakowski, 1962; Alexander and Bigelow, 1960; Bush, 1966, 1969a; Askew, 1968; Maynard Smith, 1966; White, 1968, 1970).

There is no question that geographic isolation can frequently be invoked to explain the evolution of new species in many sexually reproducing animals including parasitic insects. The inferred sequence of events involving the physical separation of populations, their subsequent genetic divergence in isolation in response to ecological changes, and eventual acquisition of reproductive isolating mechanisms is well known. There are ample, albeit reconstructed, examples to support the view that the allopatric model of speciation is the major if not exclusive means of speciation in all sexually reproducing animals (Dobzhansky, 1970; Mayr, 1963).

The concept itself is straightforward. Although the evidence in support of the allopatric model of speciation is based on rather subjective interpretations of the origin and evolution of existing species and by some increasingly shaky support from theoretical population genetics, the trend has been to accept the simplistic model. It is argued without really objective evidence that changes in assortative mating patterns, habitat selection, and associative learning behavior (induction or conditioning) are seldom if ever under the control of single genes. Models of sympatric speciation, which frequently are based on the assumption that these factors may be controlled by only a few genes, are therefore dismissed as unrealistic. Furthermore, speciation, we are told, can occur only after the accumulation of many co-adapted genetic changes generated in the absence of any disruptive effects of gene flow between diverging populations.

It is also widely accepted that speciation necessarily involves a genetic revolution and a lengthy time period. Mayr (1954 and later papers) has gathered an impressive array of examples to support this view. Unfortunately, none of these examples present empirical evidence at the genetic level that has been collected from natural populations actually undergoing speciation. The degree of genetic divergence is always estimated from species long after the speciation event has occurred. What is being measured is not the amount of genetic differentiation necessary to insure reproductive isolation (in other words speciation) but the amount of evolution that has occurred during and following the speciation event. A genetic revolution may indeed occur between species, but the genes involved may have little to do with reproductive isolation.

There is actually a growing body of evidence that speciation may occur rapidly without a genetic revolution or long period of geographic isolation. In nature distinct sibling species or semi-species of insects have evolved within short periods of time. For example, several new species of the pyraustid moth genus, *Hedylepta*, have radiated on banana from an originally coconut-infesting form since the bananas were introduced by man to the Hawaiian Islands one to two thousand years ago (Zimmerman, 1960). The rapid establishment of re-productive isolation between a recently established Bogota, Columbia, population of *Drosophila pseudoobscura* and the North and Central American populations (i.e. from various localities of the United States and Guatemala) has been reported by Prakash (1972). This reproductive isolation has evolved in the absence of apparent extensive genetic differentiation in the geographic isolate, suggesting that a major adaptive reorganization of the genome is not always a prerequisite for establishing reproductive isolation.

Laboratory experiments have also demonstrated how rapidly reproductive isolation can be established in insects. Under conditions of disruptive selection, Knight, *et al.* (1956) labeled two *Drosophila melanogaster* populations homo-zygous for different recessive marker genes and allowed these to mate at random. The progeny of cross mating were then eliminated while offspring resulting from the mating of like with like were allowed to mate at random. Within 15 generations a considerable, though not absolute, barrier to cross breeding had been built up. These results were taken a step further in the disruptive selection experiments of Thoday and Gibson (1962, 1970), who were able to obtain complete reproductive isolation in *D. melanogaster* by selecting for extremes in sternal-pleural chaetae number within 12 generations under sympatric conditions between two lines of flies originating from only two females. Lack of extensive genetic differentiation has also been found in orchids (van der Pijl and Dodson, 1966) where even intergenetic hybrids may be compatible. Reproductive isolation is maintained by specific insect pollinators. Furthermore, Dobzhansky and Pavlovsky (1966) and Dobzhansky (1972) have noted the spontaneous origin of an incipient species in the *D. paulistorum* complex within a few generations under laboratory conditions. As in the *D. pseudoobscura* example already mentioned, these studies suggest that major genetic alterations of the genome are not always necessary for speciation.

The rapid sympatric origin of new host races and species is not only supported by field observation and experimental evidence but by theoretical models as well. The work of several authors has demonstrated that genotypes which utilize several 'sub-niches' can be added to a population even if they are of inferior viability to the normal 'niche' of the species (Ludwig, 1950; Levene, 1953; Basykin, 1965). However, these models assume random mating and do not take into account the possibility of certain genotypes moving preferentially to 'niches' for which they are best fitted, or that there may be a tendency for mating to occur within a 'niche' rather than at random. Such attributes would greatly enhance the formation of host races under sympatric conditions. Maynard Smith (1966) has taken these factors into account and shown that under disruptive selection a stable polymorphism, once established, may lead to sympatric speciation whenever there is a positive correlation between mate and 'niche' selection. This is exactly the case in those parasitic insects that mate on the host plant or animal (Bush, 1966, 1969a; Askew, 1968).

Thus there is strong circumstantial evidence that in some insects, particularly in parasitic forms, reproductively isolated host races can rapidly develop sympatrically in nature and in a short time evolve into distinct species. There is, however, no complete analysis at the genetic level of any actively evolving host race complex to evaluate the importance of host shifts in the speciation process. Information is almost entirely lacking on basic questions concerning: (1) the role of host plant induction at the genetic level on host selection, (2) the genetic basis of host selection, and (3) the effects of gene flow on the local differentiation of natural populations and the speciation process in parasitic forms.

This situation undoubtedly reflects in part the difficulty of working with parasitic insects and also the lack of opportunity and techniques for studying the genetic basis of incipient host race formation and speciation as it develops in nature. The recent appearance of new host races in certain economic pests of domesticated food plants coupled with new technological developments, such as gel electrophoresis, to study genetic variation in natural populations has dramatically changed this situation.

These new host races on domesticated plants of agricultural importance provide excellent opportunities to study the process of host race formation and speciation in natural populations. Several native North American species of *Rhagoletis*, for instance, have developed new host races on introduced European fruits such as apples and cherries (Christenson and Foote, 1960; Bush, 1966, 1969a). For a number of reasons which will be discussed shortly, these flies offer a unique combination of biological attributes that make them a most ideal group of insects for studies on speciation mechanisms. Therefore, in 1961 I undertook a long term study of the genus and its host races.

Evidence gathered during the early stages of this study led to the development of a tentative model of sympatric speciation based on genetic changes in host preference and emergence patterns (Bush, 1969a). However, certain essential pieces of information to substantiate various assumptions on which this model was based were lacking. Over the last three years an effort has been made to obtain information on various behavioral and ecological aspects of these flies and their relatives and to examine the genetic basis of host selection. The results have helped to clarify many points and placed the model on a firmer basis.

LIFE HISTORY AND HOST PLANT RELATIONSHIPS

To establish valid assumptions on which a realistic genetic model of host race formation and speciation can be based by necessity requires specific information on certain interrelated aspects of the ecology, behavior and genetics of the parasite in question. For example, allochronic isolation is one factor that can limit the rate of gene flow between host races adapted to plants with different fruiting times. An understanding of the factors involved in the synchronization of emergence and adult activity such as diapause and environmental cues is essential.

Although diapause is facultative and dependent upon certain light and temperature regimes (Prokopy, 1968a) there is normally only one generation per year in almost all temperate climate species of *Rhagoletis*. A small second brood is produced in newly established host races in some years on introduced domesticated fruits that mature earlier in the summer than fruit on the original native host plants. In the new hosts the larvae are exposed to day lengths to which they have not yet become fully adapted. These second brood adults seldom mature in time to oviposit as they emerge very late in the season (Porter, 1928). The appearance of the second brood in these newly established races suggests that diapause is under genetic control as is the case in several insects (Danilevskii, 1965; Morris and Fulton, 1970).

The adults eclose from over-wintering pupae at various times during the summer, depending on the species or race. Emergence, mating and oviposition activity generally coincide with the time of maximum availability of host fruit suitable for oviposition. In native fruit this period is relatively short compared to the fruiting times of domesticated fruits, such as cherries and apples, where artificial selection has produced varieties that mature much earlier or later than their wild counterparts.

The establishment of populations on new host plants has tended to alter the emergence time of species like the apple maggot (*R. pomonella*) which originally infested only hawthorn as well as the cherry fruit flies *R. cingulata*, *R. indifferens* and *R. fausta* (Bush, 1966, 1969a). Also, a shift to one host plant has sometimes provided a bridge between plants that do not fruit at the same time of year in natural populations.

Such is the case with the new cherry race of *R. pomonella* discovered only eight years ago (Shervis, *et al.*, 1970) in Door County, Wisconsin. The vast majority of the cherry fruits infested by the cherry race mature and fall from the trees long before the fruits of the original hawthorn race are available for oviposition. The gap in fruiting time is filled by the apple race which appeared on apples in the Hudson River Valley in 1864 and rapidly spread over most of the eastern apple growing areas of North America (Bush, 1969a). It is now associated primarily with moderately early maturing domesticated varieties, such as Red Astrachans, that are in prime condition for oviposition approximately midway between the fruiting times of cherries and haws in the Door Peninsula.

The highly localized cherry race of *R. pomonella* appears to have arisen in response to changing agricultural practices. In Door County, Wisconsin, cherries frequently are not picked because of a declining market. Therefore fruit remains on the tree in abandoned, unsprayed orchards much longer than in the past and provides a new oviposition resource for early emerging apple maggot adults.

Although the emergence of certain species such as the apple maggot may extend from late June (in cherries) to October (in hawthorn) in Door County, Wisconsin, the average life span of an individual is considerably less than the total three month activity period. Some data suggest that the mean is between 20 to 30 days in *R. pomonella* under natural conditions (Porter, 1928), although this has never been established with accuracy in any *Rhagoletis* species. It is clear, however, that few flies infesting cherries would live long enough to oviposit in hawthorns.

Dispersal in *Rhagoletis* is dependent on the presence or absence of host fruit in the area. If abundant fruit of a susceptible variety is available, there appears to be little inclination for the flies to leave the trees. However, if the fruit crop fails in any given year, which is a common phenomenon in alternate bearing wild host plants, then the flies may move considerable distances to other areas where fruit is available. *R. completa* and *R. pomonella*, for example, have been recovered up to one mile from their release points (Barnes, 1959; Maxwell and Parsons, 1968), while *R. pomonella* adults tagged under natural conditions with Strontium 90 at a single feeding station in an apple orchard with abundant fruit were found to cover only a short distance (400 yds) from the labeling site (Neilson, 1971).

HOST SELECTION AND MATING BEHAVIOR AS REPRODUCTIVE ISOLATING MECHANISMS

A *Rhagoletis* fly uses several long distance physical and chemical cues to locate its host plant (Fig. 1, Step 1). Using Bird Tanglefoot coated plywood and plastic panels of different colors, shapes, and patterns we have found that in the absence of chemical cues adult *R. pomonella* and *R. fausta* of both sexes orient to any large yellow or dark colored (red, blue or black) mass (Moericke, *et al.*, 1973). The panel is rendered even more attractive to the flies if the color is presented as a double checkerboard on two layers of clear plastic. This apparently simulates the leafy structure of a normal tree. The yellow color used in our studies reflects much light in the area of the spectrum reflected from the leaf's surface (Prokopy, 1972b). The dark colors probably represent reflectance characteristics of trees when approached by the fly from the shady side. These physical cues are not host specific, but are simply used by the flies to locate a potential host plant of the right size, shape, pattern, and color. As such they cannot be regarded as critical isolating mechanisms. The more important host identifying cues are chemical in nature.

The principal olfactory cue thus far demonstrated in *pomonella* is fruit odor (Prokopy *et al.*, 1973). At what distance from the plant the flies perceive the odor is unknown. Not all domesticated apple varieties, however, are equally attractive, which may in part explain why some apple varieties are less susceptible to attack than others. Native seedling host trees would possess greater variation in fruit odor within a species than domesticated varieties of grafted isogenic stocks. This is an important point to remember when the mechanism of host race formation is considered below.

Once on the tree (Fig. 1, Step 2) a 'leaf factor' holds *R. pomonella* on a suitable host plant longer than on a non-host plant or stimulates both sexes to look more extensively for host fruits (Bush *et al.*, 1973). The fruits are located by strictly visual cues such as size, shape, and color (Fig. 1, Step 3) (Prokopy, 1968b, 1969; Prokopy *et al.*, 1971). When a fly arrives on a fruit, it moves slowly over its surface apparently measuring its shape, size, color, and surface texture (Prokopy, 1966, 1967, Wiesmann, 1937; Prokopy and Boller, 1970). Preliminary experiments suggest that it also may pick up some chemical cue(s) (Bush, unpublished).

Each *Rhagoletis* species has a preferred range of fruit size. *Rhagoletis cornivora, R. zephyria,* and *R. cerasi* select small fruit (ca. 10 mm diameter) for oviposition; *mendax* chooses slightly larger fruit (ca. 10-20 mm diameter); and *pomonella,* which differs from all the others, accepts a wider range of fruit sizes, but prefers laying eggs in fruit of 20-40 mm diameter (Wiesmann, 1937; Prokopy and Boller, 1970; Prokopy and Bush, 1973b). There is an indication that the hawthorn and cherry races of *R. pomonella* may prefer slightly smaller fruits than the apple race (Prokopy and Bush, 1973b).

Although the chemical cues are important isolating mechanisms, neither the leaf factor nor fruit odor is in itself an absolute prerequisite for oviposition. Males and females will visit red wooden spheres lacking any fruit odor hanging in non-host trees such as poplar and birch. When apples are placed in these

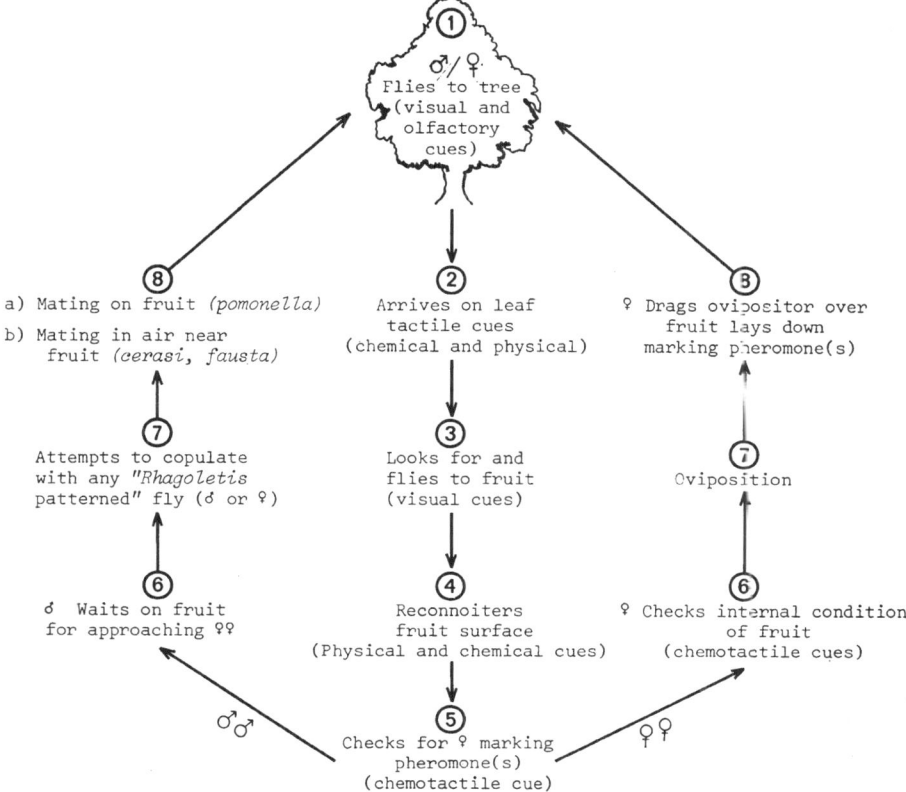

Figure 1. Host selection and mating behavior of *Rhagoletis*.

trees eggs are laid, but in smaller numbers than when hung in adjacent apple trees from which all other fruit has been removed (Bush *et al.*, 1973). Whether females would oviposit in fruits other than apple or cherries has not been established. Females, however, could occasionally ovipost in non-host fruits if the correct physical and chemical cues were provided by the fruit itself. Glasgow (1933), for instance, found *Rhagoletis cingulata* ovipositing or attempting to oviposit in non-host fruit in which larvae could not survive.

The female inserts her eggs under the skin of the fruit and at this time determines the internal condition of the fruit (Fig. 1, Step 6). Again, physical and chemical cues are probably involved in oviposition but their importance and functions are not yet determined. Caged adult females will lay in artificial hollow wax domes (Prokopy and Boller, 1970) and in artificial fruits made only of agar covered with a thin parafilm® skin of the proper size (Boller, 1968) which suggests that the external and internal chemical and physical cues may not be absolutely essential in eliciting egg deposition.

After an egg is deposited, the female drags her ovipositor around the fruit (Fig. 1, Step 8), laying down one or more marking pheromones that inhibit oviposition by other females and also holds males on the fruit or in the area where females are active (Prokopy, 1972a; Prokopy and Bush, 1972).

The male responds to the same cues as the female up to the point of oviposition (Fig. 1, Steps 1-5). From this point on, the male devotes almost all of his time on the fruit to mating behavior (Fig. 1, Steps 6-8). Unlike many insects, in *Rhagoletis* and probably most *Tephritidae* in the sub-families Trypetinae, mating occurs at the oviposition site. The host plant thus acts as a rendezvous site for courtship and mating. This critical factor in the biology of these flies has an important bearing on their ability to establish new host races and on the problem of speciation in parasitic insects in general as many of these animals mate on their host plants or animals.

Visual cues such as body coloration and wing pattern are important in *Rhagoletis* courtship behavior (Bush, 1966, 1969a; Prokopy and Bush, 1973a), but they cannot be regarded as important reproductive isolating mechanisms in sibling species groups such as the *pomonella* group that have always shifted to new hosts in the course of speciation. Because mating occurs on the fruits of different hosts these species never encounter one another during the crucial period of courtship and mating. Only in species groups that have radiated on the same host plant by geographic speciation and have established sympatric colonies, as in the *R. suavis* group which infest only walnuts, are visual cues important isolating mechanisms (Bush, 1969b).

Although we have never observed *Rhagoletis* mating anywhere other than on their host fruit(s) on occasion a male and female of different races or species might encounter one another on a fruit growing on a non-host tree. Visual cues in this case could reduce the likelihood of interspecific mating if body markings are strikingly different. They do not completely inhibit attempts of interspecific mating as we have observed *Rhagoletis fausta* and *R. pomonella in copula* on cherries in Door County, Wisconsin, (Prokopy and Bush, 1973a) where both species now infest the same fruit. These matings, however, result in almost no viable eggs (Bush and Prokopy, 1973a).

Long distance sex attractants are known to be important in some tephritid species (Christenson and Foote, 1960). These could reduce the importance of host specificity in reducing gene flow between host races. However, it has not been possible to demonstrate any long distance sex attractants in *Rhagoletis* (Prokopy and Bush, 1972).

In species of the *pomonella* group, all of which infest unrelated fruits, the most important reproductive isolating mechanism is therefore ecological in nature, namely that of host plant preference. It is for this reason that the chemical cues on the one hand and possibly to a lesser extent fruit shape, size, color, and texture on the other are normally the most important factors in limiting gene flow between newly established host races in sibling species (Bush, 1969a).

HOST SHIFTS AND THE GENETIC REVOLUTION

Most models of sympatric speciation assume that a genetic revolution is *not* a prerequisite for reproductive isolation to evolve. There is evidence that this is true in those species of *Rhagoletis* that have recently established new host races or which have shifted to a new host in the course of speciation.

Bush and Prokopy (1973b) have found that in the recently established cherry race of *R. pomonella* there has been no drastic shift in gene frequencies at two polymorphic loci (PGM and IDH) from the frequencies found in the apple race from which it was derived. There are in fact only slight quantitative differences from Nova Scotia to Wisconsin between sympatric populations of the apple race established over 100 years ago (Bush, 1966, 1969a) and the parental hawthorn race. Preliminary data on three other polymorphic loci (GOT, PGI and ADH) have given identical patterns of variation (Bush, unpublished). No electrophoretically detectable differences have been noted at six other monomorphic loci.

In a parallel study on the cherry (*Prunus*) and honeysuckle (*Lonicera*) host races of the European cherry fruit fly, *Rhagoletis cerasi,* the same lack of genetic variation has been found (Bush and Boller, 1973). Although the cherry and honeysuckle races appear to be much older than the more recently established apple and cherry races of *R. pomonella,* they cannot be distinguished on the basis of morphology or karyotype. The two races and their respective host plants occur in close sympatry over most of their ranges in Europe.

Hybridization tests between and within host races have established two general patterns of differentiation in these flies (Boller and Bush, 1973). There is considerable incompatibility (as measured by egg viability) between eastern and western populations of the cherry infesting race. In all crosses involving western males and eastern females almost no viable eggs were laid. These two populations have all the characteristics of the so-called 'semispecies' of the *paulistorum* complex in *Drosophila* (Dobzhansky, 1972). Although reproductive isolation is strongly developed between the eastern and western representatives of this species, no unique alleles or striking frequency differences are found in five polymorphic loci (PGM, PGI, GOT, IDH and ADH) in samples taken from 24 widely separated localities reaching from Spain east to Turkey and from Holland south to southern Italy. They were identically monomorphic at seven other loci.

As there is evidence of strong genetic incompatibility between the eastern and western populations of *cerasi* on cherries, gene flow cannot account for the similarity in alleles and frequencies. Indeed there are now rather compelling reasons to believe that, contrary to conventional wisdom, gene flow may in some cases have only limited influence on differentiation along environmental gradients and between host races (Ehrlich and Raven, 1969; Antonovics, 1971; Endler, 1973). In the case of *R. cerasi* it is obvious that the two populations have evolved considerable isolation in the complete absence of any apparent genetic revolution.

A second pattern of differentiation exists between the cherry and honeysuckle races. Hybridization experiments have established that sympatric populations of the two races are not as strongly reproductively isolated from one another as the eastern and western populations of the cherry race. Furthermore, no significant differences in frequencies or alleles could be found between the races in two out of three localities (Czechoslovakia and Hungary). At the third locality in Germany, an allele is present at low frequency in the honeysuckle race that is absent in the sympatric cherry infesting population. The same allele, however, does occur at low frequencies in other cherry infesting populations in other countries. There is also a slight but significant difference in the frequency of the shared alleles at two of the four most polymorphic loci in flies in Germany.

As in the case of the eastern and western cherry races, no genetic revolution appears to have occurred or to be underway between the cherry and honeysuckle races although the two populations infest host fruits of unrelated families and therefore have very different ecologies and behavior patterns and have probably been isolated for a considerable period of time. Selection rather than gene flow must account for the similarities in gene frequencies.

THE GENETICS OF HOST SELECTION AND SURVIVAL

One of the most basic assumptions in any model of sympatric host race formation is that host selection and survival are ultimately under genetic control. What is the evidence in tephritids and other insects to support such an assumption? An underlying problem also concerns the complexity of the genetic basis of host selection itself. Can host preference be modified as a result of a single mutation or is host selection polygenic in nature and thus requires the alteration of many loci before a permanent shift in host preference can occur?

Clearly the entire process of host and mate selection involves a series of complex interactions between the insect and its environment as outlined in Figure 2. Therefore, host selection must be under the control of many genetic loci as rightly emphasized by Mayr (1947, 1963). Does it necessarily follow, however, that it is impossible for a single mutation to alter host selection or survival and have a meaningful effect on gene flow? As pointed out earlier, only a few of these insect-host plant interactions are actually critical to specific host plant identification. These cues in *Rhagoletis* are primarily chemical in nature and they are monitored by the insect through contact and olfactory chemoreceptors.

Some insect chemoreceptors are odor and taste specialists such as the larval inositol receptor of the medial sensillum of *Manduca sexta* which responds only to inositol (Schoonhoven, 1969a). Others are odor and taste generalists which may respond to either a particular class of compounds such as the amino acid receptor in *Pieris brassicae* larvae (Schoonhoven, 1969b) or even a broader range of chemical attractants and deterrents. The type of receptors present in an individual will naturally vary between species depending on their host specificity and host seeking behavior. A species with a broad host range perhaps may have a more diverse repertoire of receptor types than a monophagous species or simply more receptor generalists than specialists.

Discrimination between host plants or between chemical cues may be based either on peripheral filtering by receptor specialists or in the case where a receptor responds to a spectrum of odors the burden of discrimination may lie with the central nervous system (Dethier, 1966; Schneider, 1969). In this case discrimination is based on the critical decoding of the total sensory pattern received from receptor generalists rather than by peripheral fine-sensory filtering.

Single gene mutations could alter the ability of a receptor specialist to recognize a particular chemical or alter the interpretation of incoming impulses from receptor generalists and change host preference from a plant previously unacceptable to an acceptable plant without altering the terminal receptors. Mutations affecting receptor specialists have not as yet been demonstrated but Ishikawa *et al.* (1963) has found a mutant of *Bombyx mori* with abnormal feeding habits which still receives the same type of information as a normal larva through the maxilla indicating that single gene mutations can alter discrimination patterns in the central nervous system.

Another factor of genetic importance concerns survival. Although the fruit utilized by most *Rhagoletis* species (with the exception of the *suavis* group which presents special problems, see Bush, 1969a and b) are all edible and probably have few oviposition or feeding deterrents, they apparently all do not serve equally well as larval substrates. Laboratory experiments have demonstrated that although *R. pomonella* can oviposit in fruits such as tomatoes that are never infested by this species under natural conditions, larval mortality is high and only a few individuals are able to develop fully and pupate. As tests were not run on adults emerging from these pupae it is not clear if these abnormal larval diets might also affect adult fecundity and viability (Pickett, 1937; Pickett and Neary, 1940; Hall, 1938). Furthermore, Glasgow, (1933) found that under natural conditions *R. cingulata* larvae failed to develop in non-host fruit growing near a cherry orchard even though eggs were deposited. All evidence therefore suggests that some genetic modifications of one or more survival genes is required before a successful shift to a new host can occur.

There is well documented evidence for the genetic control of resistance by antibiosis in plants to insect attack (Beck, 1965). These factors for the most part act on the larvae although they could have effects on adult fecundity, viability and host seeking behavior as well. Survival of the parasite frequently depends on genes which can counter the antibiotic activity of the plant. These genes are generally unrelated to those associated with host selection. Unfortunately, the genetics of survival in parasites is as yet poorly understood as most research has focused attention on plant resistance.

The few studies that are available indicate that an alteration of a single locus can drastically affect survival. In the Hessian fly, Hatchett and Gallun (1970) have shown that the ability of Race A to survive in the resistant wheat cultivar Seneca and of Race E to survive on Monon are controlled by single recessive gene pairs at different loci. In both races or 'biotypes' there is a gene-for-gene insect-host plant relationship with each resistant gene in wheat having a complementary gene for survival in the insect. Similar gene-for-gene relationships between cultivars and parasitic biotypes have been found in aphids (Briggs, 1965) and in the potato nematode, *Heterodera rostochiensis* (Jones and Parrott, 1965). Coevolution for resistance and survival genes between host plants and their parasites is probably a common phenomenon in nature.

In monophagous insects mutations at only one or two loci might be sufficient to insure survival on a closely related new host. More extensive genetic changes with respect to survival may be needed when more radical host shifts are involved. This may explain why the host plants utilized by oligophagous insects are frequently closely related, and fall within the same species group, genus or tribe. Polyphagous species which may feed on plants of unrelated families appear to have a more diverse array of detoxifying mechanisms such as a higher level of mixed function oxidases (Feeny, 1971).

The only evidence for genetic changes affecting host selection between sibling species of Tephritidae is the recent study of Huettel and Bush (1972). The genetic basis for host selection was tentatively established by a series of hybrid crosses between two sympatric species of essentially monophagous gall-forming *Procecidochares*. Two very closely related species or conspecific Compositae of the genus *Heterotheca* are infested by *P. australis* and another undescribed *Procecidochares* (species A) infests *Macroanthera phyllocephala*. Both host plants, which belong to the tribe Asterae, grow together along the Gulf Coast of Texas. In these flies a single locus apparently controls the ability to discriminate between the two hosts and other Compositae. The mechanism of host selection is yet unknown, but appears to involve token stimuli of a chemical nature.

When homozygous for a particular 'host recognition' allele only one host plant is selected. Furthermore, forced oviposition experiments have established that larvae of *P. australis* cannot survive in *M. phyllocephala* and species A cannot develop on *Heterotheca*. Both plants can serve as larval hosts for F_1 hybrids, but normally when offered a choice F_1 adults deposit almost all their eggs in the host plant of the P_1 parent (the host in which they were reared). This suggests that some form of strong associative learning by induction acts on the F_1 progeny as all F_1 individuals share equally the genes of both P_1 parents and should select either host plant at random for oviposition which is obviously not the case.

Back crossed progeny, on the other hand, behave quite differently. About 50 per cent of the B_1 females oviposit on the host plant *on which they were not reared*. The remainder oviposit on both plants or only on the host plant on which they fed as larvae. The F_1 hybrid and backcross results indicate that although there is a primary gene with alternate alleles affecting host plant selection, there is also a genetic component that determines the ability to be induced to oviposit on one or the other host plant. Individuals homozygous for one of the major 'host recognition' alleles cannot be induced to the other host plant irrespective of which induction alleles are present simply because these individuals lack the ability to recognize it. In heterozygous individuals, on the other hand, the gene or genes affecting induction can be expressed as both major host recognition alleles are present. The intensity of induction in these heterozygous backcross individuals would depend on the number and dosage of induction alleles present. These would be evenly balanced in F_1 hybrids so that individuals could be strongly induced to either host plant. In *Procecidochares* the degree of induction was high (96-98 per cent). This result was not surprising for a monophagous species which is undoubtedly highly canalized genetically for genes associated with host recognition, survival and induction. In B_1 individuals induction genes would not be in balance and induction in these individuals was not as tight as in the F_1 hybrids. Furthermore, the presence of the proper 'host selection' allele in at least the heterozygous condition is required before the effects of induction can even be expressed.

Similar but less rigidly controlled effects of induction have been found by Jermy *et al.* (1968), who examined the effects of host plant conditioning on the larvae of the tobacco worm, *Manduca sexta*, and the corn ear worm, *Heliothus zea*. Larvae of these species could be induced only on the plants which fall within their normal host range represented in each case by members of closely related species or genera. This also appears to be the case in the egg parasite *Trichogramma simifunatum* (Taylor and Stern, 1971). Manning (1967) and Herschberger and Smith (1967) following up Thorpe's studies in 1939 on *Drosophila* have demonstrated that adult oviposition behavior could be altered by exposure in the larval diet to the taste and odor of peppermint to which it is rarely ever exposed in nature.

On the basis of our current understanding of insect-host interactions there appears to be at least three more or less discrete major genetic components involved in the adaptation of an insect to a new host. These are genes involved with *host recognition, survival,* and *induction.* All of these genetic factors must therefore be taken into consideration in developing any model of sympatric speciation.

The most important of these in *Rhagoletis* with respect to shifts to new host plants appears to be the host recognition genes. Survival could be important in monophagous and oliphagous species when shifts are made between hosts with widely different chemistries. Induction genes, on the other hand, appear to have the least effect during the initial stage of colonization, but selection acting on these genes could be important after the shift has been made to improve on the efficiency of host recognition.

A MODEL OF SYMPATRIC HOST RACE FORMATION AND SPECIATION IN RHAGOLETIS

There are two major prerequisites essential for a successful shift to a new host plant: (1) the shift must occur in an area where both the old and new host plants occur together, (2) in frugivorous insects the fruiting times of the two hosts must also overlap. Rarely, if ever, would a successful shift occur allopatrically or allochronically as a result of a chance long distance dispersal by a fertilized female or by some early or late emerging fly.

The chance that a few individual males or females after dispersing to a previously unoccupied area or time period would have the genetic constitution suitable for recognizing, ovipositing and surviving in a new host plant is small. As dispersal in *Rhagoletis* usually occurs before the female reaches sexual maturity (5-12 days) it is even more unlikely that she would have mated. This means that a male of the right genotype emerging at the same time must also be attracted to the same isolated tree and fruit if the female is to be fertilized.

On the new host plant the opportunity for encounters between individual flies with the proper genetic makeup would be enhanced in areas where the two host plants occur together in large numbers in close proximity and with at least overlapping fruiting times. Also, contact would permit repeated experiments at the site of colonization to be made with various gene combinations until the correct combinations of host-seeking and survival alleles were brought together in individual flies. Host shifts are therefore most frequently initiated and new colonies established under sympatric conditions in oligophagous and monophagous Tephritidae. With this fact in mind we may now look at one possible sequence of steps that might lead to host race formation and speciation.

The model presented here is based on the following assumptions which appear quite realistic in light of recent experimental evidence. 1. Diapause and emergence times are ultimately under genetic control. 2. Orientation to and selection of a host plant is in response to a chemical cue. 3. Host selection (i.e., the behavioral response to the primary token stimulus) is controlled by a single locus. H_1H_1 individuals recognize only the original host A; H_1H_2 can recognize both hosts but by induction (See 5 below) move preferentially to host A, while H_2H_2 individuals can recognize only the new host B. 4. Survival in the old and new host is controlled by alleles at a single locus. S_1S_1 and S_2S_2 individuals can survive only on host A and B respectively; S_1S_2 individuals can survive on both host plants. 5. The range of host plants to which individuals can be induced (conditioned) is under polygenic control. The level of canalization in host preference would depend on the number of plant species in the normal repertoire of the insect. Seldom if ever would their effect on host selection be absolute. 6. Mating occurs on the host plant. A less detailed model based on similar assumptions was presented by Bush in 1969a.

That natural populations of insects sustain a large reservoir of genetic variation is now well established (Lewontin and Hubby, 1966; Johnson, *et al.*, 1966). Mutations affecting host selection, survival and induction would undoubtedly have ample opportunity to arise in the large populations of *Rhagoletis* frequently generated in areas of fruit abundance.

These new alleles could be lost through drift, retained in low frequencies if closely linked to highly beneficial genes or held at intermediate frequencies by overdominance even in situations where the allele is lethal in the homozygous condition. Lethals or semi-lethals of this type are well known in several natural populations including man (Cavalli-Sforza and Bodmer, 1971; Mayr, 1963).

First we will consider host race formation that involves a shift to a fruit that is not too unlike the original host. The most critical first step in this case is for a mutation to occur that establishes a new 'host recognition' allele ($H_1 \rightarrow H_2$). If the new allele is not lost by chance and it becomes established in a small area, eventually recombination will produce individuals homozygous for the new allele. These individuals will be attracted to the new host plant B if it occurs in the area. H_1H_2 flies would be induced by the old host A and rarely if ever move to or oviposit on the new host B. Because some larvae might survive without a change (at least initially) in survival genes, a small founding population of H_2H_2 flies would be established that returns in subsequent generations for mating and oviposition on the new host plant B. Mutations at one or more loci improving survival might occur later.

The establishment of a new host race through the modification of a single allele is similar in some respects to the so-called 'hopeful monster' type of systemic mutation that results simultaneously in reproductive isolation and ecological compatability once proposed by Goldschmidt (1948) and now largely discredited. However, a mutation resulting in a host shift is clearly no different than any other type of adaptive mutation and there is no need to erect a special class of 'existential adaptations' as Goldschmidt did to account for speciation.

If the new host plant is very different and if the hybrid larvae cannot survive or develop poorly in the new host, then mutations would have to occur at both a 'host recognition' locus and at an appropriate survival locus before a successful shift can occur. The new allele at the survival locus would confer some protection to the larvae from the antibiotic activities of the plant. Such a mutation could involve simply a modification in the activity or amount of an enzyme already present much like many mutations that confer resistance to pesticides such as DDT (Oppenoouth, 1965; Georghiou, 1973) and naturally occurring toxicants (Smith, 1962). The new survival allele could be held in the parent population at low or intermediate frequency just as the 'host recognition' alleles are maintained. Some individuals in the original host A population will eventually be produced that are either heterozygous for S_1S_2 or homozygous S_2S_2 and also homozygous for H_2H_2. These individuals would mate and produce progeny on the new host plant B that can survive and develop into normal adults. Once established on the new host any S_1S_1 progeny produced through recombination would be essentially homozygous for a functional lethal and would die. Selection would therefore strongly favor the S_2S_2 homozygotes. This model is similar to the two loci model of sympatric speciation proposed by Maynard Smith (1966).

It is possible to construct more complicated models involving two or more host recognition genes and multiple survival genes. Although the frequency of a successful colonization attempt is reduced with each additional essential gene, the likelihood that such a shift would lead to greatly reduced gene flow between the host races is increased as are the chances for speciation.

Once a foothold has been gained the new host race would rapidly increase in numbers, particularly if the new host fruit is unexploited by other *Rhagoletis* species or fruit eating insects. During the early stages of colonization there also may be at least a temporary relaxation from parasites as was the case with the introduction of *Procecidochares utilis* from Mexico into Australia in 1952 (Dodd, 1961). This would be particularly true with respect to those parasites that find their insect host by first homing in on the prey's host plant (Read, *et al.*, 1970). A change in the parasites' host-seeking behavior would be necessary before they could locate their former prey in the new host plant.

Colonization success would depend also on the genetic composition of the new host plants. It is not just coincidence that the best examples of contemporary host race formation involve economic pests on cultivated crops. In the process of domestication, man has selected out many of the qualities of the original plant that provided protection against herbivores and parasites. Their function has been replaced by insecticides, mitecides, and fungicides. Man has also relied heavily on grafting and cuttings to propagate preferred varieties. In so doing he has essentially distributed plants sharing identical genotypes over thousands of acres.

Each variety is planted in high density monocultures, thus facilitating the colonization and spread of a new host race. Wild plants are generally more widely dispersed and genetically heterogeneous perhaps making it more difficult for a host race to get established.

The recent shift of the apple maggot in the Door County Peninsula of Wisconsin from apples to cherries is a good example. The Montmoracy cherry variety (*Prunus cerasus*) is grown almost exclusively throughout Door County and adjacent areas. Although this variety is grafted onto seedling root stalk, the trunk, branches, leaves and fruit are genetically exactly the same from tree to tree. Under these circumstances the *pomonella* population had only to establish itself on one abandoned tree and in so doing was suitably adapted for infesting all domesticated cherries of the same variety in the area. A change in a single survival gene like those occurring in biotypes of the Hessian fly could insure survival on the new host. If, as already outlined, this change in a survival gene is accompanied by a change in one or two host seeking genes, a new host race and possibly a new species is on its way.

Successful shifts to domesticated plants therefore undoubtedly must occur much more frequently than would be expected under more natural conditions. The host plants of the sibling species *R. pomonella* and *R. mendax* which appear to have speciated sympatrically on hawthorns and blueberries may serve as an example (Bush, 1969a).

Individual plants or small clones (some hawthorn species are apomictic and blueberries may spread by roots) would differ genetically from one another and a gene (or genes) that insures survival in one plant may not be sufficient to cope with slightly different factors of antibiosis present in other plants of the same species. As a result a colonization would only establish a beach head on a single plant, clone, or colony of closely related plants. The rate of population growth of the new host race would be limited and its distribution contained by the natural genetic variation of the new host plant as well as its spotty distribution.

Adaptation would necessarily be a slow process and success in establishing a permanent population and expanding into new areas would depend on many aspects of the ecology of the founding population and the inherent diversity of its resistance to antibiotic factors in the plant. Also, if the fruit crop fails for several years the host race may die out. Only on very rare occasions would colonizers possess an adequate array of genotypes to overcome the natural battery of defense mechanisms present in the new host plant species. The greater the variation in defensive mechanisms in the plants the smaller the chances of a successful colonization and speciation event.

Thus it would appear that even in phytophagous insects and other parasites, sympatric speciation would be a rare event under natural conditions. It would, however, probably be more common than speciation through the process of geographic isolation which would still require the shift to a new host plant under sympatric conditions followed by the isolation of the two host plants from one another.

Man's intervention has only simplified the process and provided a natural laboratory to study the mechanisms of speciation in these animals. The new host races on domesticated plants, although of considerable economic importance, are also probably the least likely to survive without man's assistance. A healthy introduction of individual variation particularly at loci involved with antibiosis would go a long way in reducing the number of successful new colonizations and diminish the importance of established host races or eliminate them altogether.

Little has been said thus far about the genetic effects of induction on host race formation. The importance of this genetic component in the speciation process is difficult to assess at this time primarily because necessary information on its mode of action is lacking. At the present state of our knowledge, induction appears to have little effect during the early stages of host race formation, but as a result of selection for more accurate host identification certain induction genes could be rapidly modified after a successful colonization in such a way as to increase the likelihood of host plant recognition in the adult. This process would only further enhance the degree of reproductive isolation between the parental population and the new host race.

Another factor that would further reduce the amount of gene flow between populations of the *R. pomonella* host races is fruiting time. Many plants fruit or flower at different times of the year thus conferring a certain degree of allochronic isolation between insect populations associated with different host plants. In Door County Peninsula the cherry population begins emergence in late June reaching a peak in July. Flies do not appear on apples in high numbers until about early to mid-August while hawthorn fruits, which mature very late, do not attract flies until early to mid-September.

What little gene flow does occur between the new host race and the parent population would be restricted to the area where the initial shift occurred and where fruiting times at least partially overlap. Also only individuals with the new alleles in the right combination would leave the old host for the new one. As the populations spread to new areas the effects of gene flow would be considerably reduced in areas outside the original colonization site. What effects the gene flow at the point of colonization would have on the final outcome would depend on how effectively and how fast natural selection could eliminate the non-adapted genes from the original parental host race. In the light of Endler's (1973) recent evaluation of the effect of gene flow on genetic divergence this problem seems less important than in the past. In actual fact no suitable studies on the effect of gene flow in natural parasite populations on the speciation process have been undertaken. Most interpretations are based on evidence supplied by purely theoretical studies using assumptions that are so oversimplified as to be useless.

Finally, the parasite may exert a strong selection pressure on the new host plant by reducing seed production. In response to the fly's attack, the plant may evolve new defense mechanisms against which the insect may evolve new defense mechanisms. This coevolution between the insect and the host would, over a long period of time, only further enhance the degree of reproductive isolation between the old and new host races. During the initial stages of speciation, however, the new host plant would be a passive partner.

CONCLUSION

It now appears likely that sympatric host race formation which has led to speciation has occurred many times in the past evolutionary history of certain species groups within the family Tephritidae and possibly in many other obligate parasitic insects of both plants and animals. The speciation event itself in these insects quite frequently involves the modification of only a few key loci concerned with host selection and survival. Genes associated with induction (conditioning) appear to have only limited effect on the speciation process itself but help to reinforce reproductive isolation between evolving host races.

A speciation event in parastic insects does not initially require a genetic revolution but extensive genetic changes may in some instances evolve rapidly after the new species is established on its new host plant, particularly if the speciation event has involved a radical change in host plant ecology. These changes, which may occur rapidly or take a long period of time depending upon how different the new host plant is from the original host, only help to reinforce reproductive isolation established by the alterations through mutation of a few host selection and survival genes.

The outline of events presented here is not meant to apply to all parasitic insects. In fact, the speciation process probably involves a series of events unique to each species group or genus. It is highly unlikely that any generalized pattern of speciation fits them all. Even so, sympatric speciation could be a common phenomenon in many parasitic insects that mate on or near their hosts. Only by careful study of recently established host races can this viewpoint be substantiated.

ACKNOWLEDGMENT

This work was supported by a grant from the National Institutes of Health (GM 15769).

REFERENCES

Alexander, R. D. and R. S. Bigelow. 1960. Allochronic speciation in field crickets, and a new species *Acheta veletis*. Evolution 14: 334-346.
Antonovics, Janis. 1971. The effects of a heterogeneous environment on the genetics of natural populations. Amer. Sci. **59**: 593-599.
Askew, R. R. 1968. Considerations on speciation in Chalcidoidea (Hymenoptera). Evolution **22**: 642-645.

Barnes, M. M. 1959. Radiotracer labeling of a natural tephritid population and flight range of the walnut husk fly. Ann. Amer. Ent. Soc. **52**: 90-92.

Basykin, A. D. 1965. On the possibility of sympatric species formation. Bull. Moscow Soc. Nat. Bio. Div. **70**: 161-165.

Beck, Stanley D. 1965. Resistance of plants to insects. Ann. Rev. Ent. **10**: 207-232.

Boller, E. 1968. An artificial oviposition device for the European cherry fruit fly, *Rhagoletis cerasi*. J. Econ. Ent. **61**: 850-852.

Boller E. F. and G. L. Bush. 1973. The population biology of the European cherry fruit fly, *Rhagoletis cerasi* L. (Diptera: Tephritidae). I. Evidence for genetic variation based on physiological parameters and hybridization experiments. In preparation.

Briggs, J. B. 1965. The distribution, abundance and genetic relationships of four strains of the rubus aphid [*Amphorophore rubi* (Kalt.)] in relation to raspberry breeding. J. Hort. Sci. **48**: 109-117.

Bush, G. L. 1966. Taxonomy, cytology and evolution of the genus *Rhagoletis* in North America (Diptera, Tephritidae). Bull. Mus. Comp. Zool. **134**: 431-562.

Bush, G. L. 1969a. Sympatric host race formation and speciation in frugivorous flies of the genus *Rhagoletis* (Diptera, Tephritidae). Evolution **23**: 237-251.

Bush, G. L. 1969b. Mating behaviour, host specificity and the ecological significance of sibling species in frugivorous flies of the genus *Rhagoletis* (Diptera, Tephritidae). Amer. Natur. **103**: 669-672.

Bush, G. L. and E. F. Boller. 1973. The population biology of the European cherry fruit fly, *Rhagoletis cerasi* L. (Diptera: Tephritidae). II. The population genetics of the cherry and honeysuckle host races. In preparation.

Bush, G. L. and R. J. Prokopy. 1973a. Post mating reproductive isolation among flies of *Rhagoletis pomonella* species group. In preparation.

Bush, G. L. and R. J. Prokopy. 1973b. Population genetics of North American *Rhagoletis*. In preparation.

Bush, G. L., R. J. Prokopy and V. Moericke. 1973. Contact leaf cues and their possible relation to fruit seeking behaviour in apple maggot flies. In preparation.

Cavalli-Sforza, L. L. and W. F. Bodmer. 1971. The Genetics of Human Populations. W. H. Freeman and Co. 995 pp.

Christenson, L. C. and R. H. Foote. 1960. Biology of fruit flies. Ann. Rev. Ent. **5**: 171-192.

Danilevskii, A. S. 1965. Photoperiodism and seasonal development of insects. Oliver and Boyd, Edinburgh and London. 283 pp.

Dethier, V. G. 1966. Feeding behaviour. In: Haskell, P. T., ed., *Insect Behavior*. Symposium No. 3. Royal Ent. Soc. London. p. 46-58.

Dobzhansky, Th. 1970. Genetics of the Evolutionary Process. Columbia Univ. Press, New York. 505 pp.

Dobzhansky, Th. 1972. Species of *Drosophila*. New excitement in an old field. Science **177**: 664-669.

Dobzhansky, Th. and O. Pavlovsky. 1966. Spontaneous origin of an incipient species in the *Drosophila paulistorum* complex. Proc. Nat. Acad. Sci. **55**: 727-733.

Dodd, A. P. 1961. Biological control of *Eupatorium adenophorum* in Queensland. Aust. J. Sci. **23**: 356-365.

Ehrlich, P. R. and P. H. Raven. 1969. Differentiation of populations. Science **165**: 1228-1232.

Endler, J. A. 1973. Gene flow and population differentiation. Science **179**: 243-250.

Feeny, P. P. 1971. Detoxication enzymes in the guts of caterpillars: and evolutionary answer to plant defenses? Science **172**: 579-581.

Georghiou, George P. 1973. The evolution of resistance to pesticides. Ann. Rev. Ecol. and Syst. 3: 133-168.

Glasgow, H. 1933. The host relations of our cherry fruit flies. J. Econ. Ent. **26**: 431-438.

Goldschmidt, R. B. 1948. Ecotype. ecospecies and macroevolution. Experientia 4: 465: 465-472.

Haldane, J. B. S. 1959. Natural selection. In P. R. Bell, ed., Darwin's Biological Work, Some Aspects Reconsidered. John Wiley and Sons, New York, pp. 101-149.

Hall, J. A. 1938. Further observations on the biology of the apple maggot (*Rhagoletis pomonella* Walsh) Rept. Ent. Soc. Ont. **69**: 53-58.

Hatchett, J. H. and R. L. Gallun. 1970. Genetics of the ability of the Hessian fly (*Mayetiola destructor*) to survive on wheats having different genes for resistance. Ann. Ent. Soc. Amer. **63**: 1400-1407.

Hershberger, W. A. and M. P. Smith. 1967. Conditioning in *Drosophila melanogaster*. Animal Behav. **15**: 259-262.

Huettel, M. D. and G. L. Bush. 1972. The genetics of host selection and its bearing on sympatric speciation in *Procecidochares* (Diptera: Tephritidae). Ent. Exp. Appl. **15**: 465-480.

Ishikawa, S., Y. Tazima and T. Hirao. 1963. Response of the chemoreceptors of maxillary sensory, hairs in a "non-preference" mutant of the silkworm. J. Seric. Sci., Tokyo **32**: 125-129.

Jermy, T., F. E. Hanson and V. G. Dethier. 1968. Induction of specific food preferences in lepidopterous larvae. Ent. Exp. Appl. **11**: 203-211.

Johnson, F. M., C. G. Kanapi, R. H. Richardson, M. R. Wheeler and W. S. Stone. 1966. An analysis of polymorphisms among isozyme loci in dark and light *Drosophila ananassae* strains from American and Western Samoa. Proc. Nat. Acad. Sci. **56**: 119-125.

Jones, F. G. W. and D. M. Parrott. 1965. The genetic relationships of pathotypes of *Heterodera rostochiensis* Woll. which reproduce on hybrid potatoes with genes for resistance. Ann. Appl. Biol. **56**: 27-36.

Knight, G. R., A. Robertson and C. H. Waddington. 1956. Selection for sexual isolation within a species. Evolution **10**: 14-22.

Levene, H. 1953. Genetic equilibrium when more than one ecological niche is available. Amer. Natur. **87**: 311-333.

Lewontin, R. C. and J. L. Hubby. 1966. A molecular approach to the study of genetic heterozygosity in natural populations of *Drosophila pseudoobscura*. Genetics **54**: 395-609

Ludwig, W. 1950. Zur Theorie der Konkurrenz. Die Annidation (Einnischung) als fünfter Evolutionsfaktor. Neue Ergeb. Probleme Zool., Klatt-Festschrift 1950, 516-537.

Manning, A. 1967. Pre-imaginal conditioning in *Drosophila*. Nature **216**: 338-340.

Maxwell, C. W. and E. C. Parsons. 1968. The recapture of marked apple maggot adults in several orchards from one release point. J. Econ. Ent. **61**: 1157-1159.

Maynard Smith, J. 1966. Sympatric speciation. Amer. Natur. **100**: 637-650.

Mayr, E. 1947. Ecological factors in speciation. Evolution **1**: 263-288.

Mayr, E. 1954. Change of genetic environment and evolution. In J. Huxley, A. C. Hardy and E. B. Ford, edd., Evolution as a Process. Allen and Unwin, London, pp. 157-180.

Mayr, E. 1963. Animal Species and Evolution. Belknap Press, Harvard. 797 pp.

Moericke, V., R. J. Prokopy and G. L. Bush. 1973. Visual cues in relation to distance attraction of apple maggot flies to host plants. In preparation.

Morris, R. F. and W. C. Fulton. 1970. Heritability of diapause intensity in *Hyphautria cunea* and correlated fitness responses. Canad. Ent. **102**: 927.

Neilson, W. T. A. 1971. Dispersal studies of a natural population of apple maggot adults. J. Econ. Ent. **64**: 648-653.

Nowakowski, J. T. 1962. Introduction to a systematic revision of the family Agromyzidae (Diptera) with some remarks on host plant selection by these flies. Anales Zoologici **20**: 68-183.

Oppenoouth, F. J. 1965. Biochemical genetics of insecticide resistance. Ann. Rev. Ent. **10**: 185-206.

Pickett, A. D. 1937. Studies on the genus Rhagoletis (Trypetidae) with special reference to *Rhagoletis pomonella* (Walsh). Can. J. Res. D. **15**: 53-75.

Pickett, A. D. and M. E. Neary. 1940. Further studies on *Rhagoletis pomonella* (Walsh). Sci. Agr. **20**: 551-556.

Porter, B. A. 1928. The apple maggot. U.S. Dept. Agr. Tech. Bull., No. 66. 58 pp.

Prakash, S. 1972. Origin of reproduction isolation in the absence of apparent genetic differentiation in a geographic isolate of *Drosophila pseudoobscura*. Genetics **72**: 143-155.

Prokopy, R. J. 1966. Artificial oviposition devices for apple maggot. J. Econ. Ent. **59**: 231-232.

Prokopy, R. J. 1967. Factors influencing effectiveness of artificial oviposition devices for apple maggot. J. Econ. Ent. **60**: 950-955.

Prokopy, R. J. 1968a. Influence of photoperiod, temperature, and food on initiation of diapause in the apple maggot. Canad. Ent. **100**: 318-329.

Prokopy, R. J. 1968b. Visual responses of apple maggot flies. *Rhagoletis pomonella* (Diptera: Tephritidae): Orchard studies. Ent. Exp. Appl. **11**: 403-422.

Prokopy, R. J. 1969. Visual responses of European cherry fruit flies, *Rhagoletis cerasi* (Diptera: Tephritidae). Polskie Pismo Ent. **33**: 540-566.

Prokopy, R. J. 1972a. Evidence for a pheromone deterring repeated oviposition in apple maggot flies. Environ. Ent. **1**: 326-332.

Prokopy, R. J. 1972b. Response of apple maggot flies to rectangles of different colors and shades. Environ Ent. **1**: 720-726.

Prokopy, R. J., E. W. Bennett and G. L. Bush. 1971. Mating behavior in *Rhagoletis pomonella* (Diptera: Tephritidae). I. Site of assembly. Canad. Ent. **103**: 1405-1409.

Prokopy, R. J. and E. F. Boller. 1970. Artificial egging system for the European cherry fruit fly. J. Econ. Ent. **63**: 1413-1417.

Prokopy, R. J. and G. L. Bush. 1972. Mating behaviour in *Rhagoletis pomonella*. III. Male aggregation in response to an arrestant. Canad. Ent. **104**: 275-283.

Prokopy, R. J. and G. L. Bush. 1973a Mating behavior in *Rhagoletis pomonella*. IV. Courtship. Canad. Ent. **105**: 873-891.

Prokopy, R. J. and G. L. Bush. 1937b. Ovipositional response to different sizes of artificial fruit by flies of *Rhagoletis pomonella* species group. Ann. Ent. Soc. Amer. **66**: 927-929.

Prokopy, R. J., V. Moericke and G. L. Bush. 1973. Attraction of apple maggot flies to odor of apples. Environ. Ent. In press.

Read, D. P., P. P. Feeny and R. B. Root. 1970. Habitat selection by the aphid parasite *Diaeretiella rapae* (Hymenoptera: Braconidae) and hyperparasite *charips brassicae* (Hymenoptera: Cynipidae). Canad. Ent. **102**: 1567-1578.

Schneider, D. 1969. Insect olfaction: deciphering system for chemical messages. Science **163**: 1031-1037.

Schoonhoven, L. M. 1969a. Gustation and food plant selection in some lepidopterous larvae. Ent. Exp. Appl. **12**: 555-564.

Schoonhoven, L. M. 1969b. Amino-acid receptor in larvae of *Pieris brassicae* (Lepidoptera). Nature **221**: 1268.

Shervis, L. H., G. M. Boush and C. F. Koval. 1970. Infestation of sour cherries by apple maggot: Confirmation of a preciously uncertain host status. J. Econ. Ent. **63**: 294-295.

Smith, H. S. 1941. Racial segregation in insect populations and its significance in applied entomology. J. Econ. Ent. **34**: 1-12.

Smith, J. N. 1962. Detoxication mechanisms. Ann. Rev. Ent. **7**: 465-480.

Taylor, T. A. and V. M. Stern. 1971. Host preference studies with the egg parasite *Trichogramma semifumatum* (Hymenoptera: Trichogammatidae). Ann. Ent. Soc. Amer. **64**: 1381-1390.

Thoday, J. M. and J. B. Gibson. 1962. Isolation by disruptive selection. Nature **193**: 1164-1166.

Thoday, J. M. and J. B. Gibson. 1970. The probability of isolation by disruptive selection. Amer. Nat. **104**: 219-230.

Thorpe, W. H. 1939. Further studies on preimaginal olfactory conditioning in insects. Proc. Roy. Soc. London, B., **127**: 424-433.

Thorpe, W. H. 1945. The evolutionary significance of habitat selection. J. Animal Ecol. **14**: 67-70.

van der Pijl, L. and C. H. Dodson, 1966. Orchid flowers, their pollination and evolution. Univ. of Miami Press, Florida. 214 pp.

Walsh, B. D. 1864. On phytophagic varieties and phytophagous species. Proc. Ent. Soc. Phila. **3**: 403-430.

Wiesmann, R. 1937. Die Orientierung der Kirschfliege, *Rhagoletis cerasi* L., bei der Eiablage Eine sinnesphysiologische Untersuchung). Landw. Jb. Schweiz. **51**: 1080-1109.

White, M. J. D. 1968. Models of speciation. Science **159**: 1065-1070.

White, M. J. D. 1970. Cytogenetics of speciation. J. Aust. Ent. Soc. **9**: 1-6.

Zimmerman, E. C. 1960. Possible evidence of rapid evolution in Hawaiian moths. Evolution **14**: 137-138.

Chromosomal evolution and speciation in *Didymuria*

ELYSSE M. CRADDOCK

*School of Biological Sciences, University of Sydney**

I. INTRODUCTION

The problem of speciation remains one of the most intriguing areas of evolutionary biology. It is now recognised that there is not just one, but perhaps many possible speciation mechanisms (Dobzhansky 1972). To a large degree, the mode of speciation that occurs in any given instance is a function of the biological and population characteristics of the group of organisms concerned. However, independently of their mode of origin, all fully differentiated species have one basic characteristic in common—i.e., their reproductive or genetic isolation from all other species. To appreciate the fundamentals of the speciation process, we should therefore try to trace the evolution of this single distinguishing feature in each of the possible speciation patterns. This will require a consideration of all the possible modes of speciation, a knowledge of the succession of stages and key events for each, the preconditions which might apply, and any particularly unique aspects. The origin of the species gap is the feature of central interest. Thus, it is most important to assess those events which lead to reproductive isolation.

The process of speciation should be interpretable in terms of the mechanisms of population genetics. Each species has its own constellation of genes, organized in a particular fashion, and protected from disruption by the genetic systems of any other species. Basically, speciation involves the division of a single continuous gene pool into two or more distinct and separate gene pools, each with a unique composition, and each with the potentiality of following thereafter uniquely independent evolutionary paths. In the very simplest terms, speciation can be viewed as a two-step process—firstly, isolation of two populations by some means (not necessarily genetic); and secondly, irreversible reinforcement of this isolation via a complex of genetic mechanisms. The various speciation mechanisms differ predominantly in the way in which the primary isolation is achieved, and to a lesser extent in the manner in which the final genetic or reproductive

* Present address: Department of Population Biology, Research School of Biological Sciences, Australian National University, Canberra A.C.T. 2600, Australia.

isolation is acquired. Traditional interpretations of the speciation process have relied on *geographical* isolation as the most likely and perhaps only mechanism responsible for initiation of species divergence (Mayr 1963; Dobzhansky 1970). *Ecological* or other forms of isolation, operating under sympatric conditions, have also been considered (Thorpe 1945; Maynard Smith 1966; Bush 1969). More recently, some attention has been directed to the possibility of *chromosomal* isolation playing a primary role in the speciation of some groups of organisms (White 1968). The situation in the phasmatid *Didymuria violescens* (Leach) provides a superb example of this form of isolation.

In the proposed chromosomal mechanism, karyotypic differentiation of populations represents the first stage of species divergence. The primary chromosomal differences provide the necessary isolating barrier which permits genetic differentiation of populations, and the subsequent acquisition of full reproductive isolation. The basis of the chromosomal isolation is as follows:— chromosome hybrids which result from matings between two chromosomally different parents in a zone of contact between two chromosomal forms will be partially sterile. In these chromosome heterozygotes, pairing and segregation of chromosomes at meiosis are somewhat irregular, resulting in a proportion of unbalanced gametes. The lowered fertility of the chromosome hybrids imposes a restriction on the free exchange of genes between the two chromosomally homozygous parental forms, which are thereby partially isolated

This paper will present some description and analysis of the chromosome variation in the phasmatid *Didymuria violescens*, and then briefly outline the mode of speciation that the data suggest. The major genetic events and the consequences of this particular speciation pattern will be discussed in relation to general interpretations of the mechanism of speciation.

II. KARYOTYPIC DIFFERENTIATION IN *Didymuria*

Didymuria violescens is a phasmatid or stick insect endemic to the coastal and montane eucalypt forests of south-eastern Australia; it extends from southern Queensland, through New South Wales and parts of Victoria, and to the Mt. Lofty Ranges in South Australia. Within this range, the species includes ten different and distinct chromosomal forms, which will be referred to as chromosome races. The races differ markedly in chromosome number, chromosome morphology and sex-chromosome mechanism (Table 1). Each race occupies a particular geographic area, within which all populations are chromosomally uniform (Figure 1). The distributions of the races are parapatric, and geographically adjacent races meet in narrow zones of overlap, which do not necessarily correspond to ecological discontinuities. The taxonomic status of the chromosome forms is somewhat equivocal at this stage: they might be considered as subspecies or semispecies by some, although they cannot be distinguished morphologically. The species *Didymuria violescens* is morphologically variable, but the morphological variation is not closely correlated with the chromosomal variation, and often the variation within a race exceeds that between races.

Table 1

Table 1. Karyotypic features of the ten chromosome races of *Didymuria violescens*.

Race	♂ sex chromosomes	No. of chromosome pairs			N.F.° (haploid)
		Long metacentric Group I	Medium metacentric Group II	Sub-acrocentric Group III	
39:40(m)	*XO*, metacentric X*	1	9	10	30
39:40(sa)	*XO*, subacrocentric X	—	7	13	27
37:38	*XO*	2	7	10	28
35:36	*XO*	3	7	8	28
31:32	*XO*	5	5	6	26
32(*XY*)	*XY*†	4	6	6	26
30(*XY*)	*XY*†	5	5	5	25
28(*XY*)w	*XY*†	5	2	7	21
28(*XY*ring)	*XY*, unequal ring bivalent	5	5	4	24
26(*XY*ring)	*XY*, unequal ring bivalent	5	5	3	23

° N.F. is the number of chromosome arms in the karyotype—the 'nombre fondamental' of Matthey (1945)

* The primitive X of XO races is metacentric. If the X form is not quoted, it is of this type.

†XY bivalents have one pairing segment: those in the 28(*XY*ring) and 26(*XY*ring) races have two pairing segments.

Diploid chromosome number ranges from 26 to 40 in females. The notation system indicates the diploid number for each race, and the sex-chromosome mechanism involved. Figure 2 diagrams the male configurations of the three kinds of mechanism found amongst the chromosome races of *Didymuria*. Five races have a single sex-chromosome in the male, and an $XO(\male):XX(\female)$ system, which is that characteristic of phasmatids (Hughes-Schrader 1959). These are the 39:40(m), 39:40(sa), 37:38, 35:36 and 31:32 races. In all but one, the X-chromosome is a large metacentric chromosome: in the 39:40(sa) race, the X is a long subacrocentric chromosome with markedly subequal arms (*cf*. Figure 3). The remaining five races have derived sex-chromosome mechanisms of the neo-$XY(\male)$: neo-$XX(\female)$ type, which incorporate some autosomal material in addition to the primitive X-chromosome. The 32(*XY*), 30(*XY*) and 28(*XY*)w races possess the same type of mechanism, defined as the primary *XY* system. This system has a longer X-chromosome, termed a neo-X, and a short acrocentric neo-Y chromosome, which pairs with a short terminal region on one arm of the X. The 28(*XY*ring) and 26(*XY*ring) races possess a more complex neo-*XY* system, defined as secondary, in which the neo-X' and neo-Y' chromosomes form an unequal ring at meiosis, as a result of pairing of both arms of the neo-Y' with the terminal regions of both arms of the neo-X' chromosome.

The amount of chromosomal variation found within the single species *Didymuria violescens* is remarkable. Interpretation of this variation in terms of

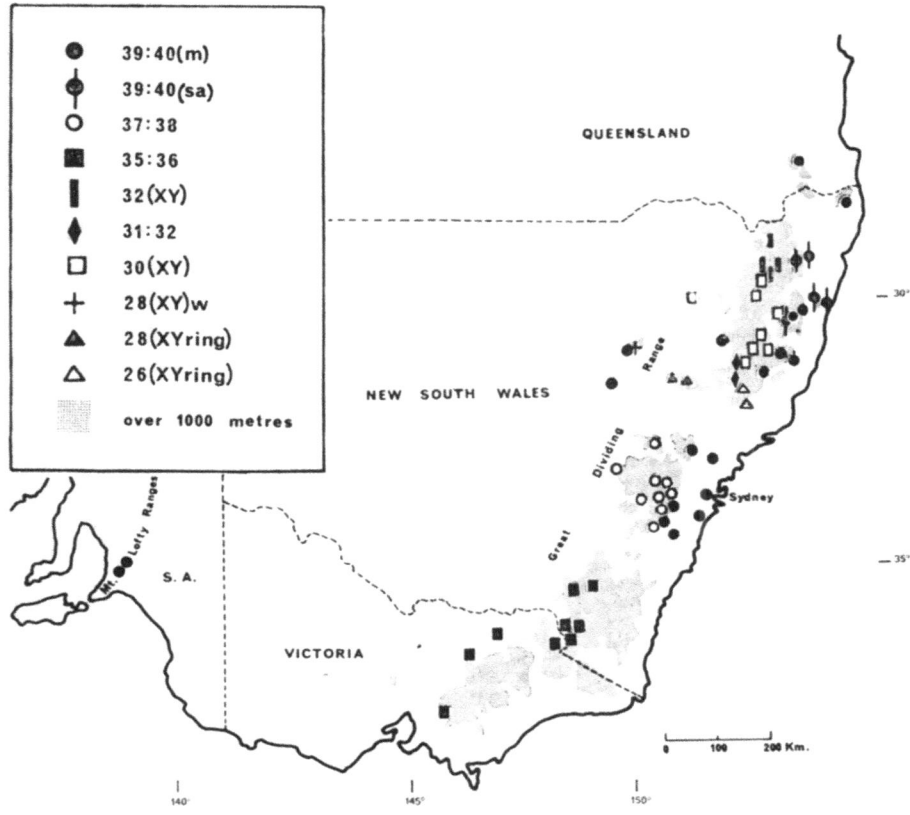

Figure 1. Distributions of the ten chromosome races of *Didymuria violescens* in south-eastern Australia.

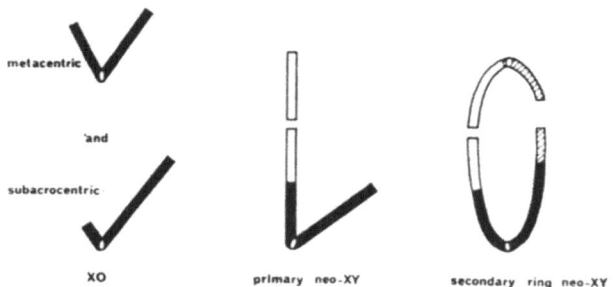

Figure 2. Male meiotic metaphase configurations of the three kinds of sex chromosome system found in *Didymuria*. The XO system has a metacentric X-chromosome in four races, and a subacrocentric X in one race.

the phyletic evolution of the species requires a knowledge of the kinds of chromosomal change involved and the structural relationships between the various karyotypes. The consequences and the significance of the observed chromosomal differentiation as a means of isolation between races are of even greater interest. If the reproductive relationships between races are such that there are genetic barriers within the species, then it will be possible to argue that racial patterns of chromosomal variation are indeed relevant as intermediate stages in one of the processes of speciation.

The chromosomal relationships

The karyotypic differences between the chromosome races of *Didymuria* are due to reorganization of the genetic material of the species in several different ways. There has been no major loss or addition of chromosomal material in the course of the evolutionary divergence of the races, since races with different chromosome numbers have the same amount of DNA (Craddock 1971). Structural rearrangement of the karyotype is presumed to have occurred repeatedly via breakage of chromosomes and reunion of the fragments in new ways. The kinds of rearrangements, and the approximate number of rearrangement events involved can be determined in two ways. Firstly, morphological comparisons of the various racial karyotypes can indicate some of the chromosomes involved in structural changes and the possible nature of these changes, should the karyotypes prove to be homologous. Secondly, analysis of the meiotic chromosome pairing patterns in interracial hybrids can provide precise evidence of most of the major rearrangement differences between parents, since chromosome pairing is usually specifically restricted to genetically homologous regions. Data from both sources has been used to interpret the chromosomal relationships between almost all the races of *Didymuria*.

In order to specify the kind of chromosomal rearrangement which differentiates two related races with different chromosome numbers, it was necessary to establish the probable direction of change. For reasons outlined elsewhere (Craddock 1970, 1971), it has been postulated that the primitive chromosomal constitution of *Didymuria* was close to that of the present 39:40(m) race, and that evolution in the species has been towards a decrease in chromosome number. Many of the rearrangements have thus been such as to involve a loss of centromeres.

Within each of the karyotypes of *Didymuria*, three major groupings of chromosomes can be distinguished. Group I consists of long metacentric chromosomes. Group II consists of medium to small metacentric and submetacentric chromosomes, and Group III includes all subacrocentric and acrocentric chromosomes (*cf.* Figures 3 and 4). A morphological comparison of the chromosome sets of all the races indicates certain trends. Along with the decrease in chromosome number, there has been general decrease in the number of acrocentric chromosomes (Group III), correlated with a general increase in the number of long metacentric chromosomes (Group I) and an increase in the relative lengths of some of these Group I chromosomes (*cf.* Table 1). In general terms, genetic material previously organized in separate chromosomes has become incorporated into the same chromosome. This kind of event has happened repeatedly in the derivation of the lower-numbered chromosome races.

Particular rearrangement events can be identified by more detailed karyotypic comparisons between the most closely related chromosome races. In many instances, the changes inferred from the morphological data have been verified from the pairing relationships observed in interracial hybrids. The meiotic data on experimentally produced chromosome hybrids are presented in detail in Craddock (1971, and in preparation). Several kinds of rearrangement have been identified. Some of these have resulted in a change in chromosome number; others have caused only a difference in karyotype morphology. The decrease in chromosome number can be attributed to three major kinds of rearrangement. Two of these involve only the autosomes; the remaining class of rearrangement involves translocations between autosomes and sex chromosomes. The rearrangements resulting in changes in chromosome morphology without change in number are mainly pericentric inversions and these have occurred in both the autosomes and the sex chromosomes of *Didymuria*. For example, at least three pericentric inversion rearrangements are required to account for the differences between the karyotypes of the 39:40(m) and 39:40(sa) races—one in the X-chromosome, and two in autosomes (Figure 3).

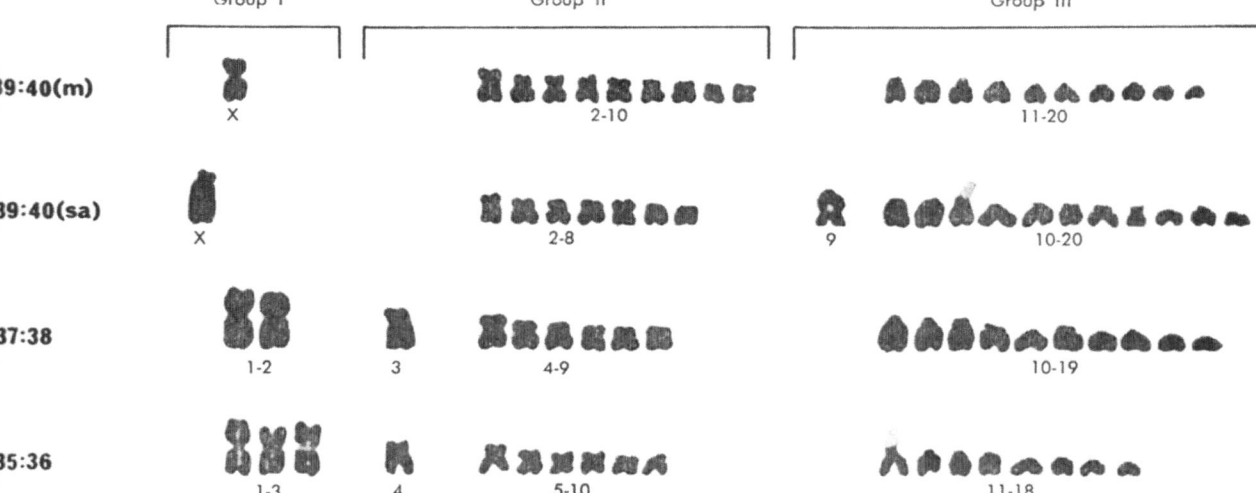

Figure 3. The haploid complements of four of the *XO* chromosome races of *Didymuria*. The karyotypes are arranged into the long metacentric chromosomes (Group I), the medium metacentric and submetacentric chromosomes (Group II), and the acrocentric and subacrocentric chromosomes (Group III). The subacrocentric X-chromosome of the 39:40(sa) race belongs to Group III, but has been displaced for comparison with the X of the 39:40(m) race.

The autosomal changes responsible for decrease in chromosome number in *Didymuria* are predominantly of the centric fusion type, and to a lesser extent of the tandem fusion type. Centric fusion in *Didymuria* has probably always been due to unequal reciprocal translocation between *acrocentric* autosomes (Darlington 1937; Tobgy 1943), rather than to strict centromeric fusion between

telocentric chromosomes. The resulting metacentric chromosome and the two
original acrocentric chromosomes, by virtue of their genetic homology, pair
together to form an association of three chromosomes (a trivalent) in the meiosis
of hybrids between races which differ by a centric fusion rearrangement. Such
rearrangements do not cause any change in the number of chromosome arms.
Tandem fusions (White 1957), which involve an unequal reciprocal translocation
between an acrocentric chromosome and a metacentric or submetacentric
chromosome, result in a reduction in the number of chromosome arms, as well
as in the number of chromosomes in the complement. Hybrids between races
which differ by a tandem fusion form meiotic trivalent associations composed of
two metacentric and one acrocentric chromosome.

   The karyotypes of the 39:40(m), 37:38, 35:36 and 31:32 races have one,
two, three and five pairs of long metacentric chromosomes respectively. The
lower-numbered XO races thus appear to have been differentiated from the
39:40(m) race via one, two and four fusion events. Male hybrid progeny from
crosses between the 39:40(m) and 35:36 races show two trivalents (plus fifteen
bivalents and the X): hybrids between the 39:40(m) and 31:32 races show four
trivalents (plus eleven bivalents and the X). Some of the fusions in this series
have apparently been accompanied by additional rearrangements such as pericentric inversions. Only in the case of the change from the 37:38 race to the
35:36 race is the increase of one metacentric pair accompanied by an exactly
corresponding decrease in the number of acrocentric chromosomes. All other
chromosomes of the two complements appear identical (Figure 3).

   Further decreases in number below the diploid number of 32 are mostly due
to translocations involving the X-chromosome (cf. below). However, the decrease
in number from the 28(XYring) to the 26(XYring) karyotype is an autosomal

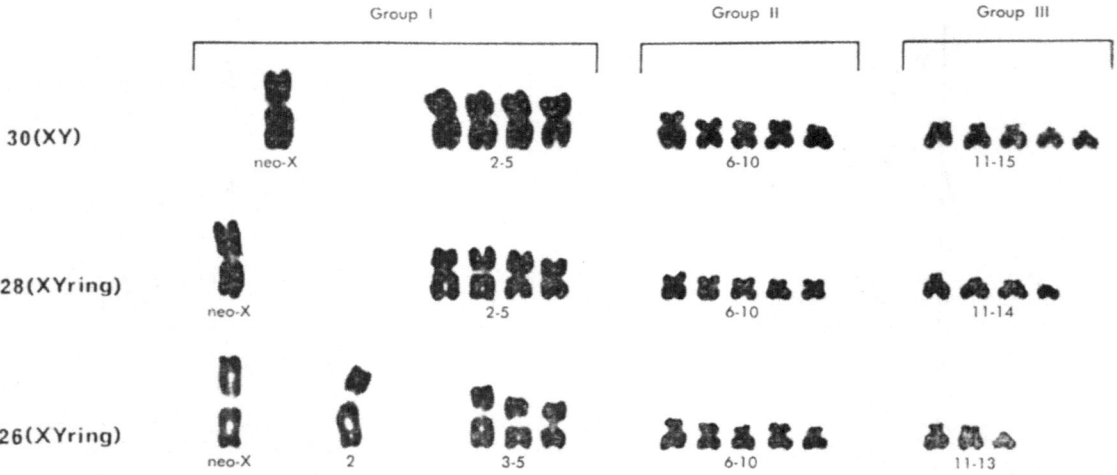

Figure 4. The haploid complements (female) of three of the XY chromosome races of
Didymuria. One has a primary neo-XY system (the 30(XY) race): two have a secondary
neo-XY mechanism.

change, probably involving a tandem fusion between a Group III acrocentric autosome and one of the long metacentric autosomes from Group I (Figure 4). A putative hybrid between these two races collected from a field hybrid zone had a single trivalent association of one acrocentric and two metacentric chromosomes, together with eleven autosomal bivalents and the XY ring sex bivalent (Craddock 1971).

The changes in the sex-chromosome system of *Didymuria* can be readily traced to translocations between autosomes and sex-chromosomes, which increase the length of the X-chromosome, as well as resulting in a decrease in total complement number. Again the rearrangements can be identified from the morphological karyotypic comparisons, and from analyses of appropriate hybrid meioses, which are given elsewhere (Craddock 1971). Table 2 presents data on the relative arm lengths of the X-chromosomes of all races. It demonstrates the following points.

Table 2

Relative arm lengths of the X-chromosomes of the races of *Didymuria*

Race	Chromosome	Mean length			Arm ratio L/S
		Long arm	Short arm	Total	
39:40(m)	Pair I	2.59±0.05	2.15±0.08	4.73±0.10	1.21
39:40(sa)	Pair I	3.91±0.06	0.93±0.03	4.84±0.08	4.21
37:38	one of pairs I and II	(2.82±0.04)	(2.32±0.03)	(5.14±0.06)	1.22
35:36	one of pairs I-III	(2.66±0.03)	(2.32±0.03)	(4.98±0.05)	1.14
31:32	one of pairs I-V	(2.71±0.04)	(2.27±0.02)	(4.98±0.06)	1.19
32(XY)	Pair I	4.05±0.06	3.03±0.10	7.08±0.15	1.34
30(XY)	Pair I	4.08±0.10	3.00±0.10	7.08±0.10	1.36
28(XY)w	one of pairs I and II	(4.13±0.03)	(2.57±0.05)	(6.70±0.08)	1.61
28(XYring)	Pair I	4.02±0.05	3.78±0.09	7.80±0.14	1.06
26(XYring)	Pair I	4.10±0.09	3.83±0.09	7.93±0.17	1.07

The mean values given in brackets refer to cases where identification of the X-chromosome was not certain. These values are therefore not as reliable.

(1) The X-chromosomes of the two 39:40 races differ significantly in the lengths of both long and short arms, but do not differ in total length. They are probably related by a single pericentric inversion.
(2) The neo-X chromosomes of the 32(XY), 30(XY) and 28(XY)w races, which all share the primary neo-XY system, are significantly longer than the X of the 39:40(m) race.
(3) The 32(XY) and 30(XY) races appear to have identical X-chromosomes, whereas that of the 28(XY)w race has probably had a separate origin. In both cases, the neo-X must have been derived from the primitive X via translocation of

the long arm of an acrocentric autosome onto one arm of the original X, thereby decreasing the chromosome number and the number of chromosome arms. In the male sex, the rearrangement remains heterozygous, and the homologous acrocentric autosome which pairs with one arm of the neo-X becomes a neo-Y chromosome.

(4) The neo-X chromosomes of the 28(XYring) and the 26(XYring) races are similar in the lengths of both arms, and in length they are longer than all other X-chromosomes. It is assumed that the ring neo-XY system has been derived but once in the history of *Didymuria*.

(5) Compared with the length of the neo-X of the 30(XY) race, the increase in length of the neo-X of the two ring races is due solely to an increase in the length of the short arm (*cf.* Figure 4). The ring neo-XY system probably originated directly from the primary neo-XY system via translocation of a small acrocentric autosome onto the short arm of the neo-X of the 30(XY) race. A further fusion apparently occurred to give rise to the small metacentric neo-Y chromosome.

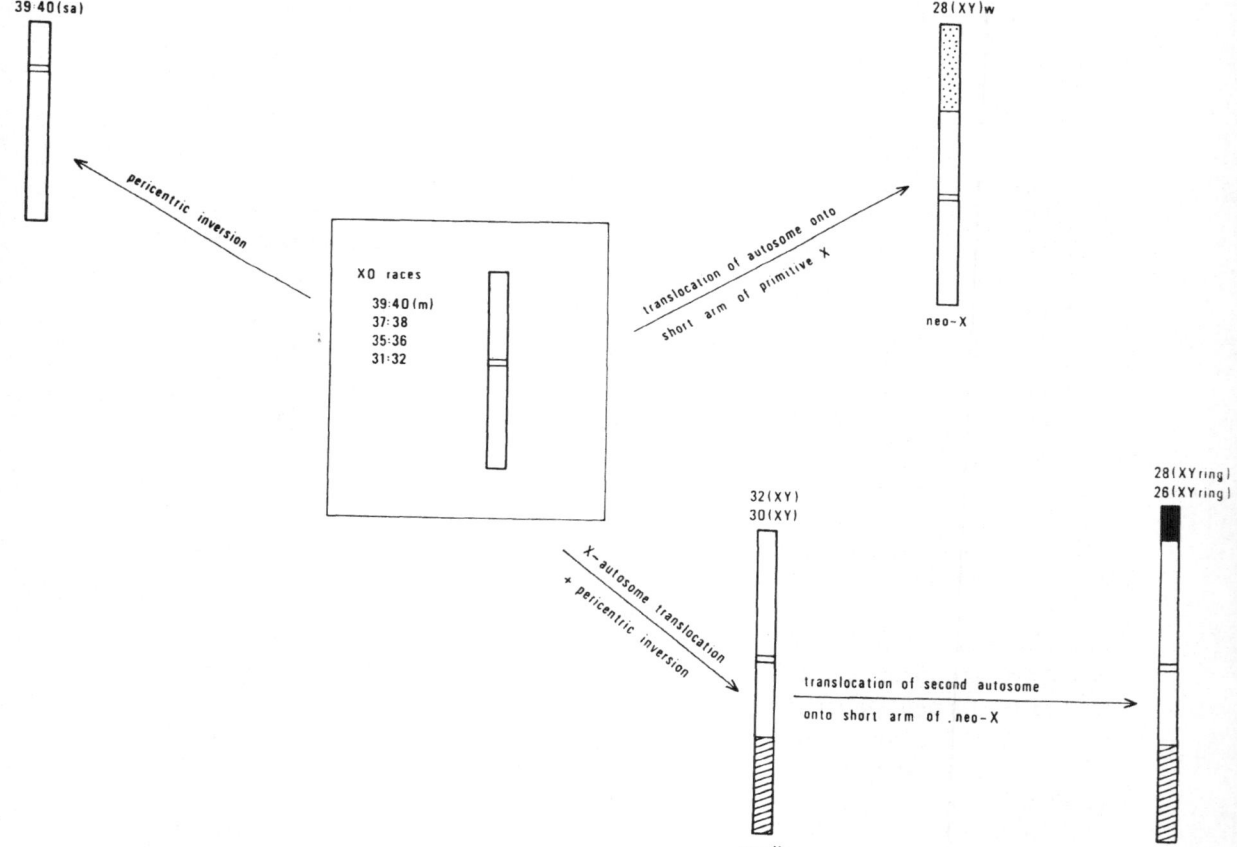

Figure 5. The postulated sequence of X-chromosome rearrangements in *Didymuria violescens*.

The relationships between the X-chromosomes of the *Didymuria* races are depicted in Figure 5. In summary, the structural changes involving the sex-chromosomes of *Didymuria* have included at least three X-autosome trans-locations, one centric fusion between neo-Y chromosomes, and at least one and probably two pericentric inversions in the X. Interpretation of all the inter-racial chromosomal relationships further suggests that the autosomal changes fixed in the species involve at least five and perhaps up to ten centric fusions,

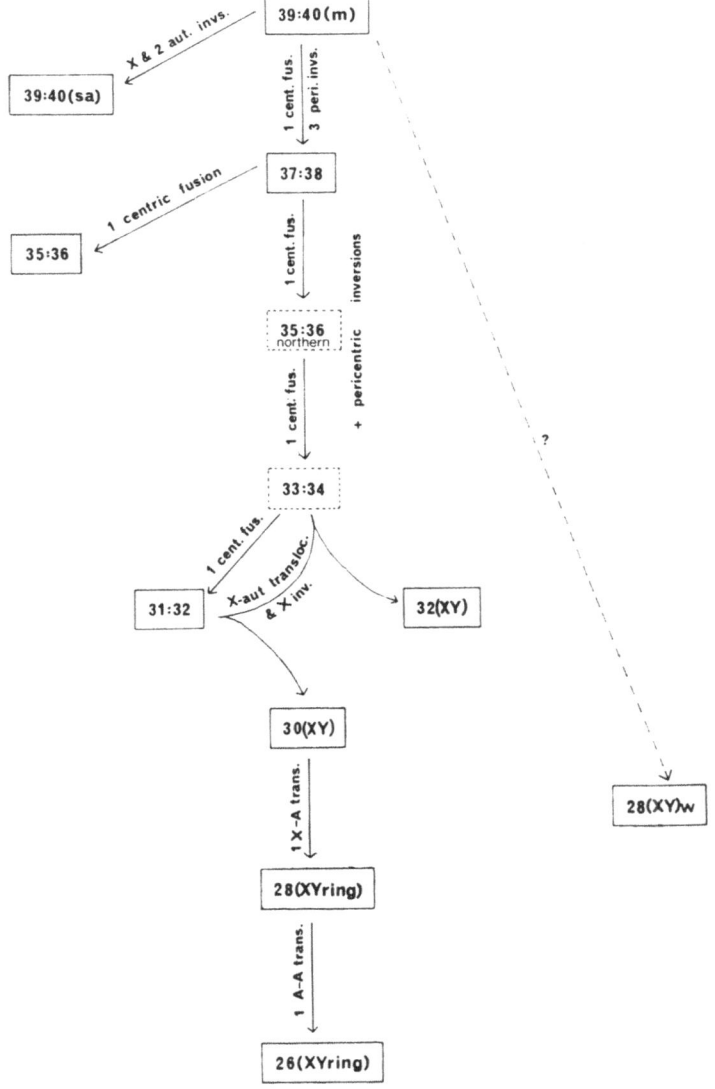

Figure 6. Scheme of the probable chromosomal interrelationships between the ten known races of *Didymuria*.

one tandem fusion and at least four pericentric inversions. This provides a minimum estimate of the structural rearrangements which have contributed to the evolutionary divergence observed in *Didymuria*. Doubt as to the origin and relationships of the 28(XY)w race prevents any more precise estimate at this stage. A scheme of the chromosomal interrelationships between the ten known races of *Didymuria* is provided in Figure 6. The interpretation draws on geographic as well as cytological data.

The reproductive relationships

In assessing the meaning of the chromosomal discontinuities between *Didymuria* populations, it is necessary to have a detailed knowledge of the reproductive events in the zones of overlap between the distributions of adjacent races. If the chromosome races are to be interpreted as incipient species, then there must be evidence of a measure of reproductive isolation between them. The degree of such isolation would depend, in the first instance, on the extent to which the chromosome differences between races restrict gene flow throughout the complex. The chromosomal barriers could represent genetic barriers if the fertility of the chromosome hybrids is reduced relative to that of chromosomally homozygous individuals.

The distributions of many *Didymuria* races are parapatric, with few geographic or ecological barriers to contact between adjacent races. Often there may be zones of overlap. In some of these zones which have been located and analysed in some detail, hybridization has been detected. The data on specific hybrid zones will not be presented here; rather, some general statements of the results will be made. Information from laboratory hybridizations will also be included where relevant.

Ethological barriers to mating between the races appear to be absent, in both the field and the laboratory. Furthermore, F_1 hybrids resulting from inter-racial matings of many chromosomal combinations are viable. The field hybrid zones contain chromosomally homozygous individuals of both parental races, as well as chromosome hybrids, and the zone of polymorphism is restricted to an extremely small area relative to the distributions of the monomorphic races. The widest known hybrid zone in *Didymuria*, which involves two races differing by a single X-autosome translocation, is of the order of four miles across. Hybrid zones between other pairs of races showing greater cytological differentiation are probably much narrower than this. The chromosome hybrids formed in field hybrid zones are not only viable, but also able to mate. The fertility of such matings is therefore the critical factor in determining the amount of reproductive or genetic isolation between chromosome races.

Lowered fertility of the chromosome hybrids is indicated by two kinds of evidence. Firstly, the restricted width of the field hybrid zones in itself suggests that the chromosome hybrids are not as fit as chromosome homozygotes. If hybrids suffered no loss in fertility, then there would be no hybrid zone at all. Rather, the polymorphism would extend throughout the entire population. The second indication that chromosome hybrids may not be completely fertile is given by the nature of the rearrangement differences which distinguish races. These are such that structural heterozygotes are very likely to undergo a degree of nondisjunction at meiosis, and so produce a proportion of chromosomally unbalanced gametes.

The fertility levels of chromosome hybrids can be partly assessed by an analysis of their cytological behaviour at meiosis. Studies of a wide range of experimentally produced *Didymuria* hybrids and some natural hybrids from field zones of overlap (Craddock 1971) have shown that their meiotic behaviour is indeed less regular than that of chromosomal homozygotes. The two main disturbances to chromosome behaviour are:—(1) a reduction in the level of chromosome pairing and (2) failure of multiple chromosome associations to orient disjunctionally at metaphase I. A third disturbance, which occurs less commonly, results from the formation of associations between nonhomologous chromosomes.

Asynapsis of homologous chromosome regions in *Didymuria* hybrids is not always confined to the chromosomes involved in structural rearrangements. Sometimes chromosome pairs which normally associate as bivalents also show a degree of asynapsis, although this is more frequent in the laboratory hybrids than in the field hybrids so far observed. The maximum single association observed involving homologous chromosomes was a trivalent. Trivalent asynapsis was sometimes partial or less often complete to give either an unequal bivalent and a univalent, or three univalents. The percentage of asynapsis in trivalents in male F_1 laboratory hybrids tested ranged from 3.3 to 77.8 per cent, with an average of 35.6 per cent. In field hybrids the asynapsis of trivalents ranged from 0 to 15 per cent between the individuals tested. The univalents resulting from asynapsis of trivalents and of bivalent associations moved randomly at anaphase, giving rise to abnormal segregations at the first meiotic division. Univalents showed other kinds of anomalous behaviour, frequently lagging on the metaphase I plate and sometimes dividing precociously. The single chromatids from these divided univalents segregated to one anaphase II pole, giving rise to unequal second division segregations. Where asynapsis was extensive early in meiosis, such cells frequently underwent restitution forming metaphase II nuclei with the diploid number of chromosomes; these nuclei subsequently produced diploid gametes.

Malorientation of trivalents on the metaphase I plate occurred in all the *Didymuria* hybrids and was equally frequent in field hybrids and experimentally produced hybrids. The linear metaphase orientation of the trivalent invariably resulted in a genetically unequal first division segregation of chromosomes, and consequently produced unbalanced gametes carrying duplications and deficiencies of the chromosomal material. Where nonhomologous associations occurred, anaphase separations similarly resulted in irregular segregations. Interracial hybrids showing these meiotic effects are thus subject to chromosomal sterility, the degree of this sterility increasing with the number of rearrangement differences between the parents. In addition to the segregational sterility, some of the experimental hybrids also showed genic sterility which was expressed in a variety of ways. (This may not be a very important contributing factor in the field hybrid zones.)

The sterility of chromosome hybrids in *Didymuria* is certainly not complete, nor would this be expected. Many of the gametes are functional and produce viable genetic and chromosomal combinations with other gametes. This has been shown by the experimental production of backcross and F_2 progenies. It has further been found that gametes carrying duplications can give rise to viable backcross progeny. Thus, in *Didymuria*, production of deficient gametes is the main cause of chromosomal sterility. Although the effective reduction in the fertility of chromosome hybrids in *Didymuria* would appear to be small, such heterozygotes are nonetheless at a selective disadvantage relative to the chromosomal homozygotes. All the currently available evidence suggests that the hybrid zones between races do indeed represent partial barriers to free gene flow throughout the species. Even though effective only at the postmating level, chromosomal translocation differences can provide a significant degree of genetic isolation.

The present partial reproductive isolation between the chromosome races of *Didymuria* would appear to provide some opportunity for genetic divergence between races and perhaps eventual specific differentiation. Because the formation of partially sterile hybrids represents some gametic wastage to the species, it might be expected that natural selection would operate to reduce the chances of interracial mating (*cf.* Dobzhansky 1940, 1970). Any mechanisms which restrict or prevent interbreeding should be strongly favoured. Thus the primary chromosomal isolation might be expected to become secondarily reinforced by isolating mechanisms which operate at the premating level. Once complete reproductive isolation had evolved, the chromosome races could attain full species status.

At present, there is little evidence of any nonchromosomal mechanisms of isolation between the races. Ethological barriers to reproduction appear to be completely lacking. All chromosome races are considered to belong to a single species, *D. violescens*. The complex of races found in this phasmatid thus provide an instance of possible incipient speciation. The races show a degree of reproductive isolation which is far from complete, but they are nonetheless potentially capable of substantial genetic differentiation and evolutionary divergence towards the level of distinct biological species.

III. THE CHROMOSOMAL MODE OF SPECIATION

The pattern of racial chromosomal differentiation observed in the phasmatid *Didymuria violescens* suggests a mode of parapatric speciation in which chromosomal barriers provide the primary isolating mechanism permitting a subsequent genetic divergence between populations. In addition to their primary isolating role, chromosomal rearrangement differences may play another important part in that they may directly stimulate the development of more efficient secondary mechanisms of reproductive isolation, via the selective forces operating in zones of contact. The basic chromosomal differentiation is thus positively involved in the initiation of speciation and in later phases of the process. This proposed mode of speciation is dependent on the special properties of major structural rearrangements. Therefore, it may have distinctive features not shared by non-chromosomal mechanisms of speciation. The major sequence of events in the chromosomal mechanism can be summarized as follows.

Firstly, a new chromosomal arrangement arises in a single individual in a population, and becomes fixed into the karyotype of all individuals of that population, displacing the existing arrangement. This is a rare event. The probability of a chromosomal mutation occurring is extremely low—much less than the probability of a gene mutation (White 1969)—since the two chromosome breaks must occur together. Further, the probability of the new rearrangement surviving, and becoming fixed in the population is even lower—it depends largely on chance events and the occurrence of random genetic drift in small populations. Once established in a local population, the new chromosome type may spread over larger territories, displacing the previously established chromosome type by a combination of chance and selective effects. The product of this spread is a chromosome race occupying a particular defined territory. Simply stated, a chromosome race is born when the occurrence of a chromosomal rearrangement coincides in space and in time with the opportunity for its establishment. This opportunity must require a very specialized set of conditions which may only be met occasionally in some species or groups of organisms, and perhaps never in other species or groups of organisms.

The completion of the speciation process is effected in the parapatric situation in the hybrid zones between adjacent races. The primary cytological barrier becomes converted to a genetic or reproductive barrier. This might involve one or other or both of the following events:

(1) a pre-mating barrier might evolve in response to selection against the formation of interracial hybrids, and/or

(2) genetic differentiation within the races might lead eventually to complete or substantial sterility or inviability of the chromosome hybrids in the zone of contact.

In this instance, mutations arising within the distributions of the various chromosome races may be arrested at the cytological barriers, thus reinforcing the partial chromosomal isolation (Key 1968). The 'tension zone' between races, which at the outset permits a certain amount of gene flow between races, may thus be progressively converted to a genetic barrier between full species. However complete reproductive isolation is finally achieved, the primary chromosomal differentiation has served as the basis for a new level of specific differentiation. Chromosomal rearrangements can thus be the prime cause of speciation in appropriate circumstances.

These basic assumptions have recently been incorporated in a model of 'stasipatric' speciation proposed by White *et al.* (1967) to account for the situation found in the *viatica* group of morabine grasshoppers. This model postulates that an essentially continuous population may be directly converted into a number of contiguous taxa, by the spread of chromosomal rearrangements around which isolating mechanisms develop. There are a number of examples of organisms with patterns of chromosomal variation involving many of the features of this model. Where the chromosomal differentiation divides the species into populations or races which are effectively genetically separate, there is every likelihood that the situation represents an initial phase of speciation via the chromosomal means. Patterns of chromosomal raciation are especially common in rodents (e.g. Wahrman *et al* 1969a, b; Thaeler 1968; Patton 1969; Nadler 1966, 1968): they also occur in certain insect groups other than phasmatids and morabine grasshoppers.

The chromosomal mechanism of speciation is not a general mode of speciation applicable in all animal groups. Although related species almost always differ karyotypically to some extent, the establishment of these differences may often be incidental to the speciation process. As already indicated, speciation via primary chromosomal differentiation may only occur under a certain special set of conditions. One of the limiting factors in the process is the chance of initial establishment of a new chromosomal rearrangement in the homozygous condition in a local population of the species. The probability of this event is strongly correlated with the breeding structure of the species and the amount of local differentiation of population which already prevails. Some division of the total population into isolates of comparatively small size seems to be a prerequisite. Probably the most important criterion is that the mobility of the organism be restricted to very low levels. The theoretical models of population genetics demonstrate that migration rate has an overwhelming effect on the possibility and extent of genetic differentiation within a total population (Wright 1969). It may be that only organisms with very restricted mobility can provide appropriate conditions for initial establishment of new rearrangements and the initiation of the sequence of stages involved in chromosomal speciation.

IV. GENETIC CONSEQUENCES OF CHROMOSOMAL SPECIATION

The genetic consequences of the chromosomal mechanism of speciation must be dependent upon the nature of the genetic events involved in the original chromosomal differentiation, and in the subsequent evolution of complete reproductive isolation. Even though it may be possible to describe, for each of the speciation mechanisms, a general sequence of stages in the development of reproductive isolation, the specific events involved in each stage need not be identical. The net genetic effect of each kind of speciation will undoubtedly differ depending on the characteristic features found in each part of the sequence. A brief comparison of geographic and chromosomal speciation with respect to the various phases in the evolution of reproductive isolation will indicate features which may be significant.

(i) The mode of origin of the primary barrier could easily affect the amount and rate of genetic differentiation possible. Establishment of a geographic barrier may be very different in its genetic effects from establishment of a chromosomal barrier.

(ii) In the subsequent stages, the effectiveness of the primary barrier in providing isolation could well influence the extent of genetic divergence finally attained. Geographic barriers may often be absolute, whereas chromosomal barriers are invariably only partial.

(iii) Different speciation mechanisms also differ in the contribution that the primary barrier makes to the development of reproductive isolation. With regular geographic speciation, the initial geographic barrier has no direct effect other than provision of favourable conditions of isolation. Chromosomal barriers, on the other hand, may make a very direct contribution to later phases of the process by their effects in stimulating the evolution of secondary reinforcing mechanisms of reproductive isolation.

Differences such as these could well affect the final form of the isolating mechanisms, the rate of speciation, and the accompanying level of genetic differentiation. It is important therefore that the genetic events involved in the speciation process be carefully and individually assessed for each particular mode of speciation, and that broad generalizations concerning the extent of genetic differences associated with species differences be avoided.

It is not possible to discuss here all the consequences of chromosomal speciation and the details of the genetic events involved at each stage of the process. Only some aspects can be considered—in particular, the involvement of genetic reorganization in speciation events of the chromosomal kind. The way in which the genetic material of each of two related species is organized may sometimes be more important than the actual number of gene differences between the species. Due to the operation of natural selection, each genetic system is an integrated one, and the way in which the balance in each system is maintained is probably unique to that system. In the chromosomal mode of speciation, genetic reorganization probably plays a very important part in at least two ways.

(1) At the level of gross organization of the genome, chromosomal rearrangement differences will have a significant effect. Transposition of segments of the genetic material to other parts of the genome may often cause chromosome regions which previously segregated independently to now segregate together. The formation of new linkage groups may in fact affect the transcription, segregation and recombination relationships of parts of the whole complement. Breakdown of existing linkage groups as a result of chromosomal rearrangement will also have important effects on the operation of differential selection. In *Didymuria*, there has clearly been a great deal of reorganization of the genetic material, via autosome-autosome and X-autosome translocations, and via inversions. Although the sequential changes might appear to be minor, the overall arrangement of the genetic material in the 26(XYring) race is certainly very different from that in the 39:40(m) race. Along with the general genotypic effects of chromosome rearrangements, there might also be some more specific effects—for example, position effects. One might ask what effect translocation of autosomes onto the X-chromosome, and incorporation into the sex-chromosome system, has on the expression of the genetic information encoded in those autosomes? Also, those translocations which involve breaks in the heterochromatic material and trans-

position of heterochromatic segments could have associated position effects, both at the interchromosomal and intrachromosomal levels.

Where centric fusions are involved, recombination effects may assume major importance. In *Didymuria,* the recombination indices of the races show a significant decrease. That in the 39:40(m) race is 39.6; in the 26(XYring) race, it is reduced to 30.1, as a result of both reduction in the haploid chromosome number and reduction in the mean chiasma frequency per cell (Craddock 1971). Changes of this order in the number of linkage groups and in recombination level will directly affect potential levels of genetic variation, so that the genetic system of the 26(XYring) race is likely to be significantly different from that of the higher-numbered races. Such changes in recombination level are generally believed to carry some adaptive significance in evolution (Lewis and John 1963).

(2) The second way in which the relationship between the process of chromosomal differentiation and genetic reorganization can be viewed relates to the effect of fixation of a new chromosomal rearrangement on the overall genetic constitution of the resulting population. This effect can be considered in terms of levels of genetic variation, and in terms of changes in coadaptation of the total gene complex, down to the level of the finer epistatic interactions. Because chromosomal mutations are unique events, each chromosomal race that now exists must have originated from a change in a single individual, and all present individuals must be descendents of that one mutant. This situation contrasts with the whole population changes usually involved in most other forms of speciation. Thus a severe bottleneck must have been imposed in the history of development of each race. The species *Didymuria* must have passed through at least a dozen such bottlenecks, with a significant loss of genetic variation occurring at each step, as a result of the involvement of small populations and the consequent inbreeding. Each of these bottlenecks can be considered as somewhat equivalent to a geographic founder event, offering the opportunity for what Mayr has termed a 'genetic revolution' (Mayr 1954). Although such bottlenecks must have been of relatively short duration, and must have been followed by some restoration of genetic variability by introgression from the parental chromosome race, nevertheless they may have far-reaching consequences. A new genetic structure may be attained by the new chromosomal type, which is substantially different from that of its ancestor. Each new chromosome arrangement may, in the course of its fixation and spread, generate a new coadapted complex, as a result of the genetic revolution and subsequent reconstruction of its gene pool under the action of selection. The genetic revolution, which would occur in any new founder population, is superimposed on the chromosomal rearrangement of the genome—a new set of interactions may have to be built up to give a workable system. In the history of *Didymuria,* numerous such complexes have been developed, as evidenced by the retention of so many of the original and intermediate chromosomal forms. Changes in the coadaptation of each new system may involve changes in various genetic control mechanisms, resulting in a serious lack of correspondence in hybrids. Chromosome pairing, for example, may be subject to different controls in the various races, so that the hybrids

might suffer serious disruption. The chromosomes of 39:40(m)—26(XY ring) F_1 laboratory hybrids of *Didymuria,* despite a high level of homology, show a very low level of pairing. These and other differences in coadaptation between races probably have some effect in restricting introgression across hybrid zones, by contributing to the level of hybrid sterility (*cf.* Key 1968).

The foregoing discussion has indicated some of the implications of species differentiation via the chromosomal mechanism suggested by studies of variation in *Didymuria* and other organisms. The involvement of genetic reorganization was emphasized because of the special circumstances and opportunities associated with the chromosomal mechanism. Still other viewpoints could be presented on the basis of the model. For a truly genetic analysis of the speciation process, predictions such as these should be tested out by independent means for this and each of the other speciation mechanisms. Only then will it be possible to compare the various mechanisms in terms of their common genetic components, and thus establish the fundamental genetic basis of the speciation process.

V. SUMMARY AND CONCLUSIONS

The phasmatid *Didymuria violescens* is geographically differentiated into at least ten chromosome races, which vary widely in diploid chromosome number, sex-chromosome mechanism and karyotype morphology. The cytological variation in the species results from the sequential fixation and spread of a series of chromosomal rearrangements, many of them fusions leading to a decrease in chromosome number. This racial pattern of karyotypic differentiation provides a basis for a degree of chromosomal isolation at the post-mating level. Because of the properties of translocation rearrangements, chromosome heterozygotes formed by hybridization between geographically adjacent races show some reduction in fertility. The narrow hybrid zones formed at the chromosomal discontinuities within *Didymuria* represent partial genetic barriers. These barriers could be significant in initiating a series of events leading to complete reproductive isolation and eventual speciation.

The interpretation of the racial pattern of chromosomal variation in *Didymuria* as an instance of incipient speciation derives some support from other similar cases showing greater degrees of reproductive isolation, and even total isolation. Speciation in certain groups of organisms showing low mobility may often proceed via the chromosomal mode, in which the establishment of rearrangement differences is the first step in speciation. The genetic consequences of this speciation pathway may be rather different from those of other mechanisms, particularly with respect to reorganization of the genome. Although speciation may occur via several different mechanisms involving different genetic events, all entail the evolution of genetic or reproductive isolation as the fundamental characteristic of the process.

REFERENCES

Bush, G. L. 1969. Sympatric host race formation and speciation in frugivorous flies of the genus *Rhagoletis*. Evolution **23**: 237-251.

Craddock, E. M. 1970. Chromosome number variation in a stick insect *Didymuria violescens* (Leach). Science, N.Y. **167**: 1380-1382.

Craddock, E. M. 1971. Cytological studies of the Australian Phasmatodea. Ph.D. Thesis, University of Sydney.

Craddock, E. M. 1972. Chromosomal diversity in the Australian Phasmatodea. Aust. J. Zool. **20**: 445-462.

Darlington, C. D. 1937. Recent advances in cytology. 2nd ed. J. & A. Churchill Ltd., London.

Dobzhansky, Th. 1940. Speciation as a stage in evolutionary divergence. Amer. Nat. **74**: 312-321.

Dobzhansky, Th. 1970. Genetics of the evolutionary process. Columbia University Press; New York and London.

Dobzhansky, Th. 1972. Species of *Drosophila*. Science **177**: 664-669.

Hughes-Schrader, S. 1959. On the cytotaxonomy of phasmids (Phasmatidae) Chromosoma **10**: 268-276.

Key, K. H. L. 1968. The concept of stasipatric speciation. Systematic Zoology **17**: 14-22.

Lewis, K. R. and B. John 1963. Chromosome marker. Churchill Ltd., London.

Matthey, R. 1945. L'évolution de la formule chromosomiale chez les vertébrés. Experientia **1**: 50-56, 78-86.

Maynard Smith, J. 1966. Sympatric speciation. Amer. Nat. **100**: 637-659.

Mayr, E. 1954. Change of genetic environment and evolution. In Evolution as a process. J. S. Huxley, A. C. Hardy & E. B. Ford, edd., pp. 157-180.

Mayr, E. 1963. Animal species and evolution. Harvard Univ. Press. Cambridge, Massachusetts.

Nadler, C. F. 1966. Chromosomes and systematics of American ground squirrels of the sub-genus *Spermophilus*. J. Mamm. **47**: 579-596.

Nadler, C. F. 1968. The chromosomes of *Spermophilus townsendi* (Rodentia: Sciuridae) and report of a new subspecies. Cytogenetics **7**: 144-157.

Patton, J. L. 1969. Chromosome evolution in the pocket-mouse *Perognathus goldmani* Osgood. Evolution **23**: 645-662.

Thaeler, C. S. Jr. 1968. Karyotypes of sixteen populations of the *Thomomys talpoides* complex of pocket gophers (Rodentia-Geomyidae). Chromosoma **25**: 172-183.

Thorpe, W. H. 1945. The evolutionary significance of habitat selection. J. Animal Ecol. **14**: 67-70.

Tobgy, H. A. 1943. A cytological study of *Crepis fuliginosa* and *C. neglecta* and their F_1 hybrid, and its bearing on the mechanism of phylogenetic reduction in chromosome number. J. Genet. **45**: 67-111.

Wahrman, J., Ruth Goitein and E. Nevo 1969a. Mole rat *Spalax*: Evolutionary significance of chromosome variation. Science **164**: 82-84.

Wahrman, J., Ruth Goitein and E. Nevo 1969b. Geographic variation of chromosome forms in *Spalax*, a subterranean mammal of restricted mobility. In Comparative Mammalian Cytogenetics. K. Benirschke, ed., Springer-Verlag, New York. pp. 30-48.

White, M. J. D. 1957. Some general problems of chromosomal evolution and speciation in animals. Survey Biol. Progr. **3**: 109-147.

White, M. J. D. 1968. Models of speciation. Science N.Y. **159**: 1065-1070.

White, M. J. D. 1969. Chromosomal rearrangements and speciation in animals. Ann. Rev. Genet. **3**: 75-98.

White, M. J. D., R. E. Blackith, R. M. Blackith and J. Cheney 1967. Cytogenetics of the *viatica* group of morabine grasshoppers I. The 'coastal' species. Aust. J. Zool. **15**: 263-302.

Wright, S. 1969. Evolution and the genetics of populations vol. 2. The theory of gene frequencies. The University of Chicago Press; Chicago and London.

Speciation in the Australian Morabine Grasshoppers — taxonomy and ecology

K. H. L. KEY

Division of Entomology, CSIRO, Box 1700, Canberra City, A.C.T. 2601, Australia

The morabine grasshoppers constitute a wholly endemic Australian subfamily of the Eumastacidae, a family with a widespread distribution, mainly in the tropics and subtropics of the Old and New Worlds. Of some 240 species of Morabinae known to exist, only 35 or 36 have been named and described. Representatives of the undescribed species are preserved in the Australian National Insect Collection in Canberra, under code numbers which have been extensively cited in published work on the cytogenetics, morphology, and biology of the group. I am at present engaged on a classification of the subfamily, employing in part numerical taxonomic methods. All the species are attenuate and completely apterous. *Moraba virgo*, otherwise a fairly typical morabine, has the distinction of being the only known obligatorily parthenogenetic species among the short-horned grasshoppers (White, Cheney, and Key 1963).

The group is found over almost the full range of temperature and rainfall experienced in Australia. Most species are found on forbs (that is, dicotyledonous herbs) and on shrubs up to head height, but about a fifth of them have adapted to grasses and a few occur on ferns. The distribution ranges of individual species are small for grasshoppers, their long axes averaging perhaps 600 km, but varying from as much as 2000 down to probably 150 or less. In Figure 1 each line represents the long axis of the putative range of a species, and the circles represent the mid-points of the lines, which are thus estimates of the centres of the ranges. It will be seen that there is a tendency for the ranges to be extended in the direction of the isohyets, which form roughly concentric rings around the continent. This indicates that rainfall is a significant factor in the distributions. Within its overall distribution, a species may be restricted by local habitat factors, especially vegetation, but restriction is often not rigid, and moreover the country beyond the limit may appear to be not different from that occupied. The critical factor is probably often some form of interaction with species having adjacent ranges, as we shall see in a moment. The number of species is greatest in moist-

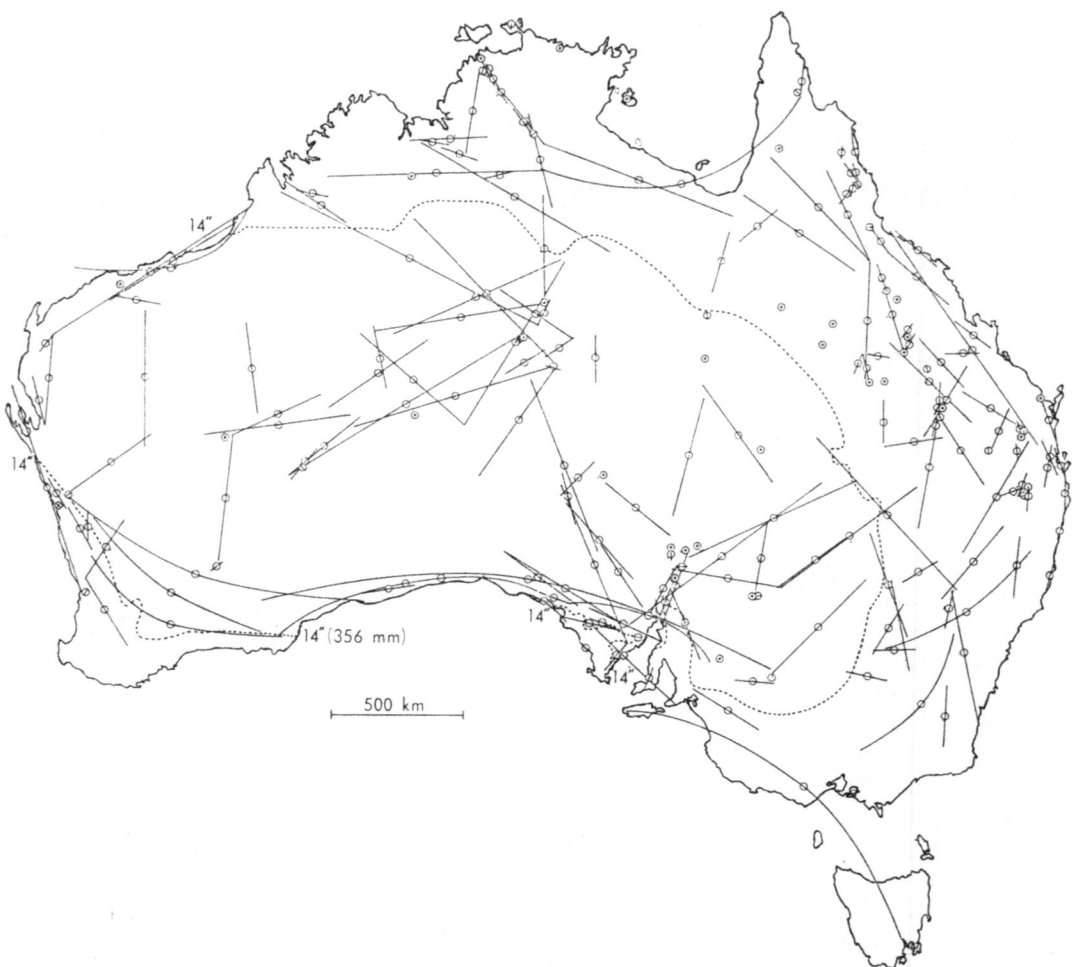

Figure 1. Long axes of the putative distribution ranges of morabine species. Circles mark the mid-points of the axes, or, when enclosing a dot, the position of the only known locality. Dotted line: 14-in. (356-mm) isohyet.

tropical and -subtropical Queensland, but it falls off in northern Cape York Peninsula and none occur in New Guinea, or in Timor.

Species of morabines in different genera commonly overlap broadly; i.e. they show typical sympatry. One must assume that they have diverged with respect to ecological niche and acquired pre-mating isolating attributes. On the other hand, closely related species in the same genus or species group often appear to be 'parapatric', to use Smith's term (Smith 1955). Their distribution ranges come into contact, but do not overlap by much more than the dispersal range of an individual within its life-time, which we may take to be at most two or three hundred metres. An apparent example of parapatric distributions

(Figure 2) is afforded by two species of the genus *Keyacris*, *K. scurra* and *K. marcida*. *K. scurra* has a more southerly distribution than *marcida*, but between

Figure 2. Distribution of *Keyacris marcida* (triangles) and *K. scurra* (solid circles, 15-chromosome race; open circles, 17-chromosome race) in south-eastern Australia.

about Orange and Cootamundra, in New South Wales, the two come into contact. The habitat of these and many other morabines has been fragmented and largely destroyed by development of the country, and particularly by sheep grazing. As a result, they are now reduced to small relict populations in cemetery reserves, along railway lines, and in other situations protected from the full impact of grazing, and it is often only by inference that one can establish the approximate limits of their original distributions. Nevertheless, in the present case both *scurra* and *marcida* have been found in the cemetery of the little town of Millthorpe, north-west of Blayney. Further south, *marcida* alone occurs in the cemetery at the village of Monteagle, 16 km north of Young, while *scurra* alone formerly occupied the cemetery at Young. At Orange there are records 14 km apart and at Cootamundra 16 km. Other examples of presumed parapatry occur in other species groups. In the genus *Warramunga* two species appear to be in parapatric contact over a distance of some 650 km.

The morabines show the same extensive intraspecific variability as is typical of grasshoppers generally—in the form of morphs, clines, and geographical races, and involving morphometrics, surface sculpture, and colour pattern. But the most striking feature of morabine intraspecific variability, and one not at all characteristic of the grasshoppers as a whole, is the widespread occurrence of chromosomal polymorphism and chromosomal races. This subject will be discussed more fully in the next contribution. Here I will be concerned mainly with its distributional, ecological, and taxonomic aspects.

Chromosomal races, which sometimes differ also in morphological features, occur in quite a number of morabine species. They are distinguished from intrapopulation polymorphisms, of course, by the fact that each chromosomal type has reached fixation over the area occupied by the race of which it may be the principal indicator, and moreover by the fact that interracial hybrids are adaptively inferior, while polymorph heterozygotes show positive heterosis. In all cases that have been sufficiently studied, the chromosomal races make parapatric contact with one another in very narrow so-called 'tension zones' (Key 1968), where intergradation occurs in their morphological characters, with more extended introgression of some, and where both chromosomal types are found, along with heterozygotes. There is no evidence of any premating reproductive barrier, and in most of the cases studied there appears to be only slight to moderate impairment of fertility in the F_1 hybrids. Nevertheless, the narrowness of the tension zone does suggest a sort of reproductive confrontation between the two races, which are continuously reinforced from the hinterland of each and in effect destroy each other in the tension zone by breeding together.

The first case of this kind to be investigated concerns again *Keyacris scurra*. This species has two chromosomal races (White 1956), the predominant and inferentially ancestral one with 15 chromosomes in the male, and a 17-chromosome race occupying a much smaller area on the western fringe of the distribution of the 15-chromosome race (Figure 2). Although the zone of contact between the races is some 200 km in length, the fragmentation of the habitat already mentioned makes it difficult to define its position with any accuracy. However, in the vicinity of Boorowa and Yass, White and Chinnick (1957) found two places (Figure 3) where the corridor between the established distributions was less than 8 km wide, and one where it was no more than 800 m. And at this last place a male with 16 chromosomes, almost certainly an interracial hybrid, was found within the population of the 15-chromosome race at its point of closest approach. White estimated from laboratory crosses that there might be a reduction in fertility of the F_1 amounting to at most 10 per cent. Some reduction in viability as well would not be surprising. An artificial population established in an unoccupied area from males of one race and virgin females of the other (White 1957) gave rise to apparently vigorous progeny and maintained itself over at least two generations, but there was no way of determining whether there may have been a small reduction in viability in comparison with a non-hybrid population. No evidence has been found of any difference between the two races in external morphology or male genitalia. The frequency of different pattern morphs and polymorphic chromosomal rearrangements seems to be the same in the two races, and the northern limit of one of these rearrangements cuts directly

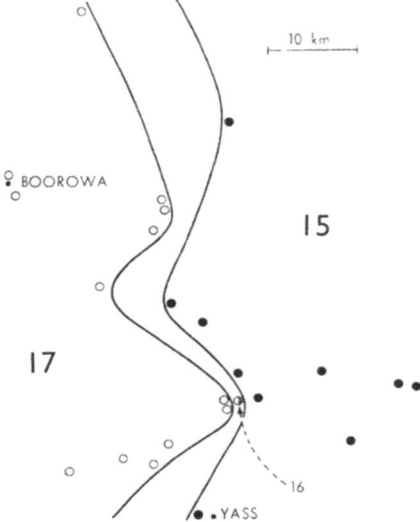

Figure 3. Distribution of the races of *Keyacris scurra* (solid circles, 15-chromosome race; open circles, 17-chromosome race) in the vicinity of Boorowa and Yass, N.S.W. Modified from White and Chinnick (1957).

across the boundary between them, all suggesting that there would have been a fairly free genetic interchange across the tension zone prior to the fragmentation of the populations during the past 100-150 years.

It will be seen from Figure 2 that the 17-chromosome race of *Keyacris scurra* occupies a lens-shaped territory between the 15-chromosome race to the south-east and *Keyacris marcida* to the north-west, and that *marcida* makes contact with the 15-chromosome race in the north, but with the 17-chromosome race in the south. If, as seems almost certain, the 17-chromosome race originated immediately to the west of the western limit of the original 15-chromosome race (White and Chinnick 1957), then it seems as though it may have expanded by pushing back the tension zones with both *marcida* and the 15-chromosome race of *scurra*. It is conceivable that this movement was still proceeding 100 years ago.

The second case of parapatric races to be studied involved a 17- and a 19-chromosome race of what is still known as '*Moraba viatica*', although this species can certainly not be retained in the genus *Moraba*. The races, commonly referred to now as '*viatica*$_{17}$' and '*viatica*$_{19}$', were found by White, Carson, and Cheney (1964) to come into contact along a typical tension zone in the heath country of south-eastern South Australia, stretching for at least 70 km between the town of Keith and the sea, and perhaps for an equal distance to the north-east In this instance thirty-three 17-chromosome individuals, along with three 19-chromosome individuals and five chromosome-number heterozygotes, were collected within a zone 400 m wide in the environs of Keith. Beyond this zone only *viatica*$_{19}$ is found to the east and *viatica*$_{17}$ to the west. It was concluded that the tension zone could hardly have been wider than 1 km and may have been less. The authors estimated that some 10 per cent of sperms produced by the heterozygotes would have been aneuploid and that viability was not greatly depressed. As in *Keyacris scurra,* there are no readily detectable external or genitalic differences between the races. Genetic interchange across the tension zone must again be inferred.

The pioneer work on *Keyacris scurra* and the 19- and 17-chromosome races of '*viatica*' was followed by the location in South Australia of typical narrow tension zones between the members of three further pairs of what I am treating as races of '*viatica*', again differing in karyotype. Moreover, it became apparent that the ten races of this species that occupy what we may call broadly 'peninsular' South Australia, and which White and his co-workers have termed the 'coastal' forms (White, Blackith, Blackith, and Cheney 1967), have apportioned the whole of this region among them, mosaic-fashion, so that no two are sympatric in the normal sense (Figure 4). All adjacent pairs can be regarded as parapatric and assumed to have exhibited the tension-zone

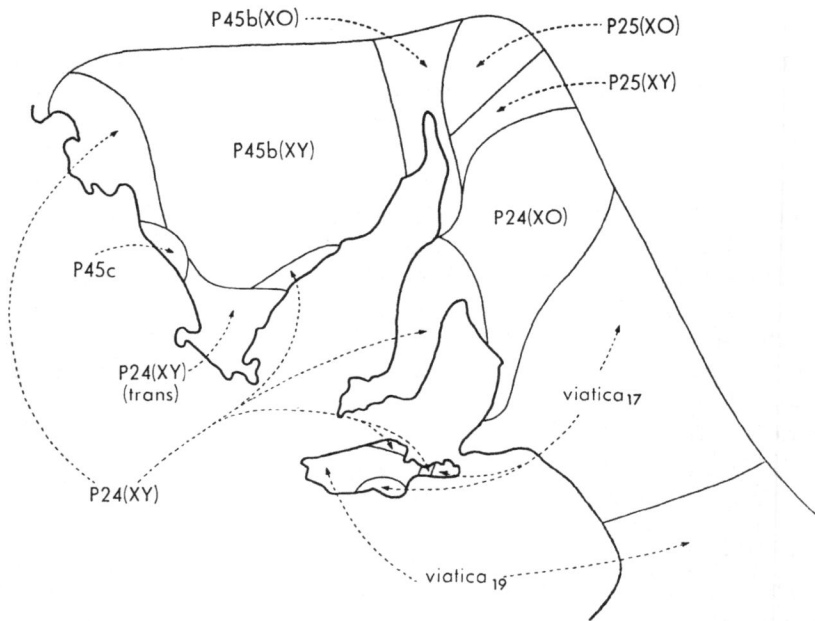

Figure 4. Mosaic distribution of the parapatric races of '*viatica*' in peninsular South Australia. Modified from Key (1968).

phenomenon at their interfaces before the European occupation. Whether relict portions of all these zones can still be discovered is another matter. A remarkable situation exists on Kangaroo Island, a large island situated off the coast of South Australia. Here White, Key, André, and Cheney (1969) found three races of '*viatica*' within a radius of 15 km, each confined to its own territory, and exhibiting between them three different tension zones (Figure 5); two were

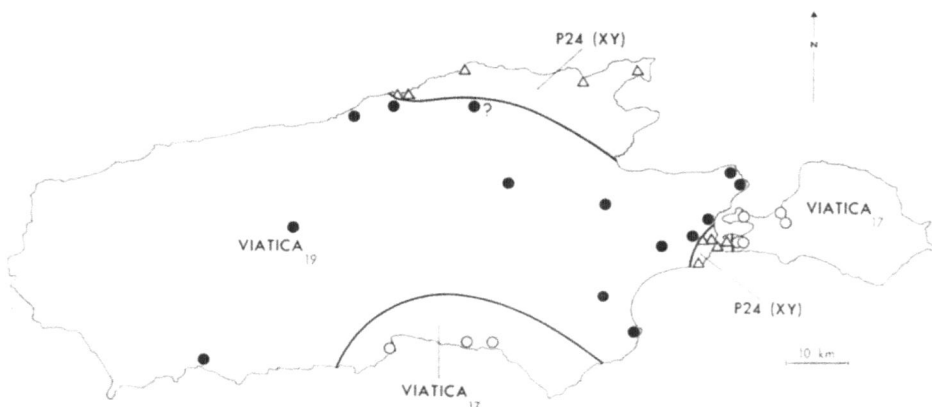

Figure 5. Distribution of three parapatric races of '*viatica*' on Kangaroo I., South Australia. Solid circles, *viatica*$_{19}$; open circles, *viatica*$_{17}$; triangles, $P24(XY)$. Modified from White, Key, André, and Cheney (1969).

successfully located, each at one point. The first of these zones, involving *viatica*$_{17}$ and the race designated $P24(XY)$, was no more than 275 m wide. Out of twenty females collected, eleven were chromosomally *viatica*$_{17}$, three were $P24(XY)$, and six were chromosomal heterozygotes. Since the racial difference involves the X chromosome, only females could be scored for chromosomal heterozygosity. The other zone, involving *viatica*$_{19}$ and $P24(XY)$, was even narrower. Although both forms were present and in immediate contact with each other, none of thirty-eight males collected showed evidence of hybridity, even though in this case not only F_1 chromosome heterozygotes, but also various kinds of backcrosses and other hybrid derivatives, would have been detectable in the male had they been present.

Typical individuals of $P24(XY)$ differ from *viatica*$_{17}$ and *viatica*$_{19}$ in certain morphological features, including the presence of a supplementary hook on the male cercus—a character that would ordinarily be regarded as of considerable taxonomic importance. Within the tension zone between $P24(XY)$ and *viatica*$_{19}$ all males could be immediately identified as one race or the other on the basis of the cercal character and there was no sign of intermediates. On the other hand, in the $P24(XY)$/*viatica*$_{17}$ tension zone many males did show intermediate cerci, thus giving evidence of a hybridity that was not detectable in the male karyotype. Moreover, some intermediacy in the cercus was traceable in the *viatica*$_{17}$ population for up to 10 km from the tension zone. A situation similar to that of the races of *viatica* is presented by two forms (originally treated as species) of the *granulosa* group of morabines, designated $P127a$ and $P127b$. To the west of Eyre Peninsula, South Australia, there is evidence that these forms, which differ in minor features of the male cercus and internal genitalia, come into parapatric contact (Figure 6) in a narrow zone in which morphological intermediates occur. In this case, however, Mr. P.-G. Fontana, in unpublished studies, has been unable to detect any chromosomal differences. Evidence from the acridid grasshoppers

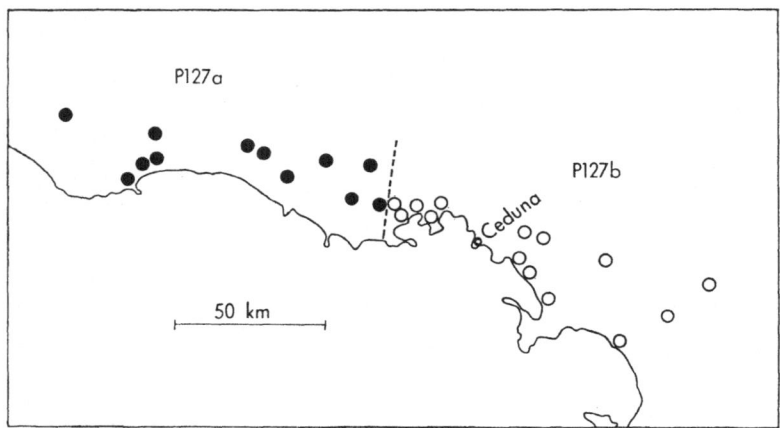

Figure 6. Distribution of two forms in the *granulosa* group of morabines, west of Eyre Peninsula, South Australia. Solid circles, *P127a*; open circles, *P127b*. Modified from P.-G. Fontana, unpublished.

and other animals also suggests that parapatry is not necessarily dependent upon structural chromosomal differences, although it may be favoured by them.

I come now to the interpretation of all this information. The simplest explanation of the tension zones would appear to be the following. A structural chromosomal rearrangement, or occasionally a major gene mutation, reaches fixation in a small geographical isolate of a species. The population of the isolate then increases and its range expands until it makes contact again with the parent population. If the resulting 'hybrids' (if one can call them that) show heterosis, or at least are at no selective disadvantage, free introgression will occur in both directions and the integrity of the species will be restored. But if the heterozygotes are at a significant selective disadvantage, as they tend to be with chromosomal rearrangements especially, the low vagility of the insects will ensure that individuals of the new form and their hybrid progeny will not be able to penetrate far into the solid phalanx of the old form before the invading gene or chromosomal rearrangement is eliminated from the lineage. The same thing will happen in the other direction, provided that the population and area of the new form are sufficiently large to 'absorb' the invaders and not be swamped by them. In this way a tension zone, or 'front' is set up, within which each form is in effect being 'controlled', and prevented from penetrating further, by reproductive confrontation with the other. However, it is really only the particular chromosome or chromosomal segment, or the particular gene, responsible for the lowered fertility or viability of the heterozygotes, whose passage is completely blocked, together with genes closely linked with these. Thus the tension zone acts like a filter, holding up some genes and chromosomal rearrangements and allowing others to introgress (Figure 7). Incidentally, the transport of genetic material may be greater than it appears, since only introgression to which some limit has been set is detectable: completely successful introgression is not.

The geographic isolate in which the critical genetic modification first

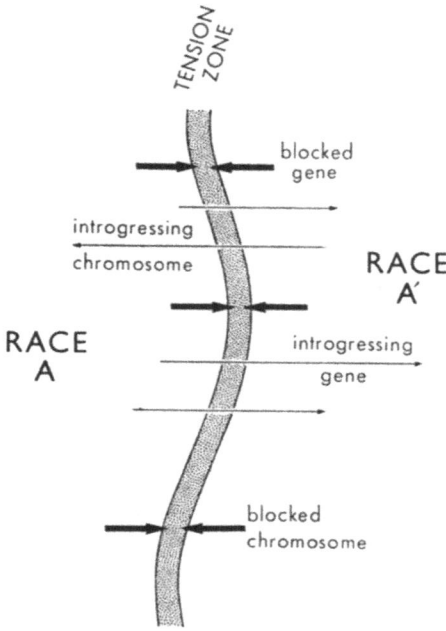

Figure 7. Diagrammatic representation of a tension zone, showing the postulated blocking of some genes and chromosome rearrangements and the introgression of others.

became fixed may be situated on the external periphery of the species distribution —that is, it may be classically allopatric; or on the kind of 'internal periphery' that may occur in animals of patchy distribution and low vagility—that is, it may be microallopatric. An external location is perhaps more likely to be effective, because the marginal conditions for survival would give greater assurance of an isolation long enough to allow the new population to expand free from contact with the parent form, and would also be likely to cause violent population fluctuations favourable to fixation. Also there is a chance that the genetic modification may prove to be pre-adapted to the more different environment on the external periphery, leading to a more rapid expansion of the population. And finally, an external location ensures that the eventual renewal of contact with the parent population will occur in one direction only; that is, the new form will not be surrounded on all sides and thus subjected to greater risk of swamping.

After the tension zone has been established, further genetic changes (whether chromosomal or genic) that may arise within one or other race (including changes that may have established tension zones of their own) will, especially if genetically linked with the material responsible for the reduced compatibility in the tension zone, tend also to be stopped there. In this way the overall genetic differences may increase and with them the intensity of selection against the hybrids. This process could lead eventually to complete hybrid infertility or inviability, as in the case of $viatica_{19}$ and $P24(XY)$ on Kangaroo

Island (which differ in three major chromosomal features as well as genically), or the case of *Keyacris scurra* and *K. marcida* in New South Wales.

This plausible, although in part still only inferential, sequence of events constitutes, in so far as it proceeds to completion, a process of speciation. It differs from the model usually invoked in that only the first step is taken allopatrically and this may involve something far short of Mayr's radical 'genetic reconstruction'. The second step is taken after contact has been resumed, but it is neither the disruptive 'semigeographic' speciation that Mayr rejects, nor the evolution of the premating isolating mechanisms of Dobzhansky, but a reinforcement of a partial postmating genetic incompatibility. It generates good biological species, between which there is no premating reproductive barrier and which remain parapatric. In the Morabinae it seems that this situation can persist for long periods of time. If in due course a premating isolating attribute is evolved, the lethal confrontation in the tension zone is broken and normal sympatry becomes possible.

So far I have said nothing about the significance of ecology in relation to the tension zones. The lines along which the parapatric races and species of morabines are separated certainly do not represent the limits of their ecological tolerance. They are far too smooth for this, as well as too narrow. Habitat diversity on both sides of the zone is much greater than any slight average difference that may exist between one side and the other, and the same spectrum of diversity is present on both sides. More central parts of the distribution areas of the two forms may or may not show appreciable differences, but these reflect minor preferences rather than limiting tolerances. A distribution determined by tolerances would yield a tremendously wide overlap in which the frequencies of the two forms would show some relation to habitat. Nor can competition, alone, account for the zones. This too would result in a broad zone of overlap (though less broad than in a tolerance-determined distribution), with a complex interdigitation and pocketing of the two forms in conformity with habitat differences. Actually the position of the zones is probably the smoothed resultant of year-to-year fluctuations of the line along which ecological conditions (perhaps mainly climatic) confer equal fitness upon the two homozygotes; but this effect is mediated by the steady reproductive confrontation, which confers great inertia upon the system.

We may suppose that at the time of the initial contact between the two forms the position of the resulting tension zone would not have fulfilled this condition. In that case it would be expected to move slowly in the direction of the less fit homozygote. This could have one of two consequences (Figure 8). Either no point of equilibrium would be attained, in which case the less adapted form might be crushed against the sea or some other ecological barrier and eliminated (there is some evidence that this may have happened, or be happening, in some cases); or, in its territorial displacements, the tension zone might reach equilibrium ecological conditions, in which case it would remain relatively static, moving only as ecological conditions changed. However, because of the inertia, superimposed upon the already very low morabine vagility, these

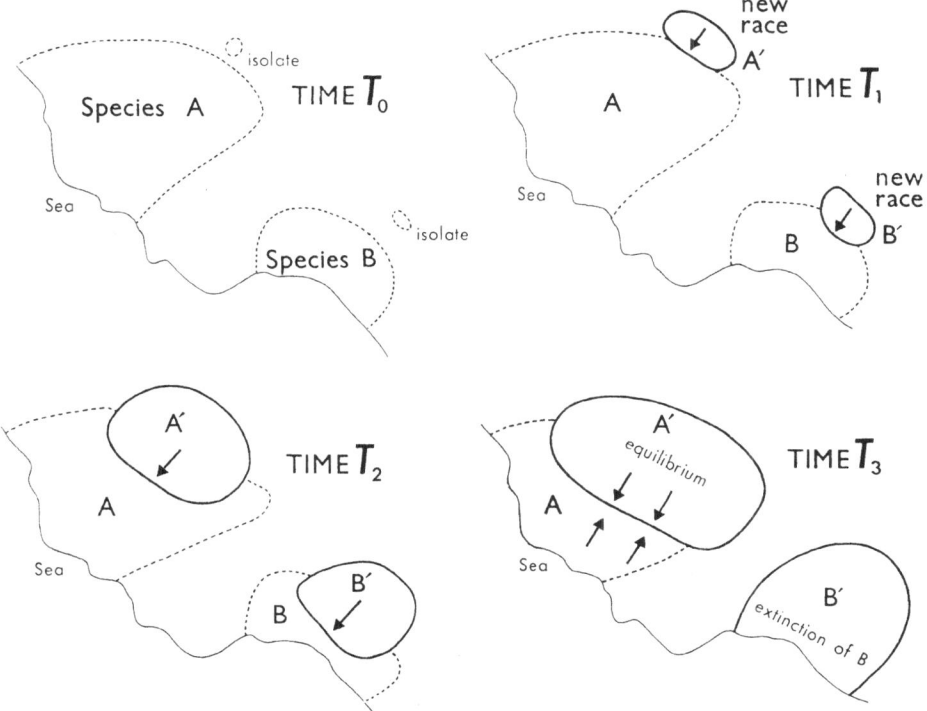

Figure 8. Postulated sequence of events in which a new race arises from a peripheral isolate in each of two species of morabines, A and B. The new races (A' and B') form tension zones with the parent populations and advance against them until (species A) 'equilibrium' ecological conditions are reached under which the two homozygotes are equally adaptive, or (species B) the parent form is eliminated.

movements are likely to be subject to a significant lag, so that the position of the zone at any time may represent a fossil adaptation to a previous position of the ecological determinant. The relative smoothness of the zone may be considered as due to the 'pinching off' of any incipient or temporary salients of one form by the partly encircling population of the other (Figure 9). The whole situation is reminiscent of war between human communities, or more especially between ant colonies. In summary, then, tension zones need not, and probably do not, indicate abrupt changes in any ecological parameter.

Speciation is the process of becoming a species. Like the process of becoming a professor, it can be identified only when it is complete. I know of no infallible way of determining in advance whether a given situation of genetic divergence and reduced gene flow is going to lead to full reproductive isolation, to introgression and reintegration, or to stalemate, as often seems to be the result in morabines; or whether perhaps one or both of the forms is going to become extinct before the question can be resolved, as is the all-too-likely prospect under the progressive domination of the earth by man. Thus, strictly speaking, speciation can only be diagnosed in retrospect.

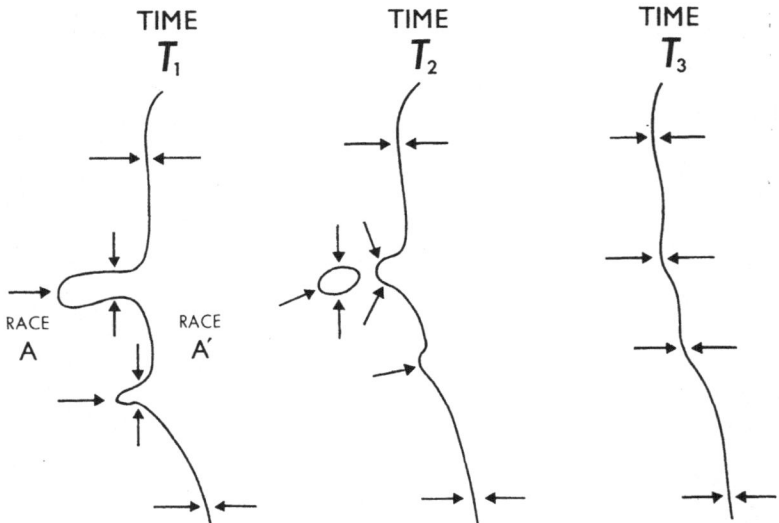

Figure 9. Tension zone between races A and A' of a morabine species, showing a postulated 'pinching off' of salients of A' by the partly encircling population of A.

A further difficulty about the concept of speciation is the difficulty with the concept of species. For bisexual animals in reproductive contact (that is, when sympatric or parapatric) this seemed to have reached a satisfactory level of definition with the publication of Mayr's concept of the biological species (Mayr 1942). Species were to be recognised as such by their reproductive isolation—by the absence of gene flow between them. Of course, in practice the state of reproductive isolation had usually to be inferred from discontinuities in morphological or other variation, and this could give rise to difficulty. Nevertheless, at least in principle the definition was unequivocal, and in principle it was susceptible of direct test. Subsequently, under the influence of Simpson (1961) and Bigelow (1965), for example, various levels of gene flow have come to be regarded as acceptable between species, provided that the 'evolutionary future' of the two forms will be separate. This places a decision essentially in the realm of guess-work. We have to arrive at some estimate of the extent of the gene flow that exists, and then we have to guess whether in all the present and future circumstances this will lead, *at some future time,* to full reproductive isolation. Not even in principle can a definition along these lines be unequivocal, or its application testable. Indeed, since under both definitions the ultimate criterion is the same, it seems peculiarly perverse to require that this condition should be met in the future, which is not open to observation, rather than in the present, which is. In any case, I believe that the percentage of presently accepted species of animals that produce, in nature, hybrids that are viable and also fertile *at all* is so very small that less harm would be done to taxonomic and nomenclatural stability by treating these as races of the same species (as the 1942 definition would require) than by treating as distinct species many of the geographic populations, now regarded as conspecific, between which there is a substantial reduction of gene flow and which may be *on the road* to reproductive isolation. This is especially so when one considers that there will inevitably be unresolvable differences of opinion as to the evolutionary future of these partially isolated forms. I therefore adopt the more rigorous earlier criterion for the biological species.

Applying this to the Morabinae, none of the ten forms already mentioned as comprising the so-called 'coastal' members of the *viatica* group should *on present evidence* be regarded as specifically distinct from *viatica*, the type specimen of which came from the Tasmanian population of $viatica_{19}$. Some of these form natural hybrids, either with $viatica_{19}$ directly, or with other races which in turn hybridise at first or second hand with $viatica_{19}$. Amongst those whose tension zones have not yet been studied or precisely located, some give rise to viable and at least partially fertile hybrids in laboratory crosses (White, Blackith, Blackith, and Cheney 1967). The remainder have not been adequately tested. In the tension zone already discussed between $P24(XY)$ and $viatica_{19}$ on Kangaroo Island, where there is no evidence of fertile hybrids, we have a case which, if it existed in isolation, would have to be interpreted as one of parapatric *species*. But $P24(XY)$ hybridises freely with $viatica_{17}$ on Kangaroo Island, and $viatica_{17}$ with $viatica_{19}$ on the mainland. Moreover $viatica_{17}$ and $viatica_{19}$ are in contact on Kangaroo Island also. Thus at least some genes must be free to pass from $P24(XY)$ to $viatica_{19}$ by the roundabout route through $viatica_{17}$, even though they cannot pass directly (Figure 10). But if $viatica_{17}$ were

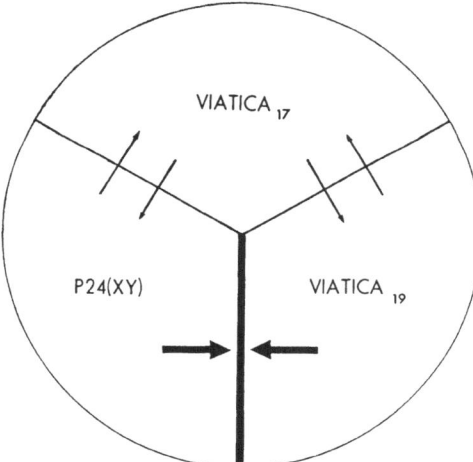

Figure 10. Diagrammatic representation of gene flow between $P24(XY)$ and $viatica_{19}$ by way of $viatica_{17}$, although direct gene flow between these forms is blocked.

to become extinct, then $P24(XY)$ and $viatica_{19}$ might well be left with no genetic communication, and the situation would be comparable with that between *Keyacris scurra* and *K. marcida*. There seems to be no doubt that the peculiar phenomenon of progressive accretion of genetic differences on tension zones does represent a mechanism (perhaps the chief mechanism) of speciation in the Morabinae, and possibly in some other groups also.

REFERENCES

Bigelow, R. S. 1965. Hybrid zones and reproductive isolation. Evolution 19: 449-58.

Key, K. H. L. 1968. The concept of stasipatric speciation. System Zool. 17: 14-22.

Mayr. E. 1942. Systematics and the Origin of Species from the viewpoint of a Zoologist. New York: Columbia Univ. Press.

Simpson, G. G. 1961. Principles of Animal Taxonomy. New York: Columbia Univ. Press.

Smith, H. M. 1955. The perspective of species. Turtox News 33: 74-7.

White, M. J. D. 1956. Adaptive chromosomal polymorphism in an Australian grasshopper. Evolution 10: 298-313.

White, M. J. D. 1957. Cytogenetics of the grasshopper *Moraba scurra*. I. Meiosis of inter-racial and interpopulation hybrids, Aust. J. Zool. 5: 285-304.

White, M. J. D., Blackith, R. E., Blackith, R. M., and Cheney, J. 1967. Cytogenetics of the *viatica* group of morabine grasshoppers. I. The 'coastal' species Aust. J. Zool. 15: 263-302.

White, M. J. D., Carson, H. L., and Cheney, J. 1964. Chromosomal races in the Australian grasshopper *Moraba viatica* in a zone of geographic overlap. Evolution 18: 417-29.

White, M. J. D., Cheney, J., and Key, K. H. L. 1963. A parthenogenetic species of grasshopper with complex structural heterozygosity (Orthoptera: Acridoidea). Aust. J. Zool. 11: 1-19.

White, M. J. D., and Chinnick, L. J. 1957. Cytogenetics of the grasshopper *Moraba scurra*. III. Distribution of the 15- and 17-chromosome races. Aust. J. Zool. 5: 338-47.

White, M. J. D., Key, K. H. L., André, M., and Cheney, J. 1969. Cytogenetics of the *viatica* group of morabine grasshoppers. II. Kangaroo Island populations. Aust. J. Zool. 17: 313-28.

Speciation in the Australian Morabine Grasshoppers — the cytogenetic evidence

M. J. D. WHITE

Department of Genetics, University of Melbourne, Parkville, Victoria 3052

Approximately 200 species of the Australian Morabine grasshoppers have been examined cytologically to some extent. Certain of the species groups or genera have been studied in depth, using a great variety of cytogenetic techniques. The majority, however, including all the tropical species, have only been subjected to a more superficial examination. Enough information exists, however, to provide material for discussion of the role of chromosomal rearrangements in the differentiation of populations, races and species. This is particularly so with regard to the major rearrangements (chromosomal fusions, dissociations, translocations and pericentric inversions in particular). Minute rearrangements, which may have been far more numerous (by one or two orders of magnitude) are very difficult to detect, individualize and quantify, in the absence of polytene chromosomes. We may obtain some information concerning them from careful chromosome measurements, especially if combined with autoradiographic studies of late-replicating DNA segments; but at the present stage our knowledge of this aspect of karyotype evolution in the morabine grasshoppers, as in other groups of organisms, is in an extremely unsatisfactory state.

The great majority of the species of morabine grasshoppers have unique karyotypes which differ from those of their closest relatives. And in numerous cases, forms which it seems necessary to regard as races of a single species have different karyotypes. In these respects Morabines are similar to many (probably most) other groups of animals. The situation which exists in a few instances in the genus *Drosophila* (particularly among the Hawaiian species), where 2, 3 or 4 related species have indistinguishable karyotypes (homosequential species-complexes) is certainly very exceptional. If in these cases there are no structural differences in the heterochromatin which have been missed by the usual polytene chromosome analysis, they demonstrate that chromosomal rearrangements are not in every instance a *sine qua non* for speciation. But because homosequential complexes are rare, even in *Drosophila*, they should not be used to argue that chromosomal rearrangements do not play an important or essential role in speciation in other or even in most cases.

In the Morabinae there are certainly instances, e.g. in the genera to which *improcera* and *granulosa* belong, where several related species appear to have indistinguishable karyotypes. But it is highly probable that in many of these cases more detailed studies (e.g. ones including determination of the DNA values, distribution of late-replicating chromosome segments and perhaps satellite DNA's) would point to the existence of hidden minor cytotaxonomic differences.

The different karyotypes which we observe in present-day species and races have, of course, arisen in the past through evolutionary fixation of chromosomal rearrangements. The great majority of the latter are undoubtedly duplications and deletions of very minute chromosome segments which are almost impossible to identify individually. In most instances these individually minute increases or decreases in the total amount of genetic material certainly involve heterochromatic segments, composed of repetitive DNA sequences of uncertain genetic function. But heterochromatin is not very conspicuous in Morabines, or at least it does not show up as strongly heteropycnotic segments, except for the X-chromosome which exhibits the usual reversal of heteropycnosis in the course of spermatogenesis (negative heteropycnosis in spermatogonial divisions, positive heteropycnosis at the prophase of the first meiotic division). However, autoradiographic studies, using tritiated thymidine have shown that one of the pairs of small autosomes is usually late-replicating in Morabines, and this may be heterochromatic and show up as a heteropycnotic mass in somatic interphase nuclei.

The major chromosomal rearrangements which are responsible for conspicuous karyotypic changes involve entire chromosome limbs or large segments of such limbs; they consequently represent a very different type of evolutionary change from the minute rearrangements. The main categories which we can identify in the Morabinae are *fusions* (decreases of chromosome number), *dissociations* (increases of chromosome number) and one type of *inversion*, the *pericentric* kind. Another type of inversion, the paracentric type, may well occur as well, but are ordinarily undetectable in grasshoppers.

From an evolutionary standpoint an important distinction seems to exist between those kinds of chromosomal rearrangements which can exist in a floating, polymorphic state in natural populations and those which can not do so. Broadly speaking, the former would be those in which the heterozygotes are heterotic or are highly adapted in at least one ecological niche of the total environment; while the latter would be ones whose heterozygotes are inferior, usually on account of some degree of reduction in fecundity. This distinction almost certainly exists in all groups of animals, although the precise types of rearrangements falling into one or the other category differ from group to group.

In the Morabinae a considerable number of species (about 12 per cent of the total) show population polymorphism for pericentric inversions, which consequently belong in some instances to the first category defined above. Such rearrangements, which seem to provide a basis for the building up of co-adapted gene-complexes in the alternative chromosomal sequences, are more likely to be cohesive than divisive factors in phylogeny. They are hence unlikely to be involved in speciation, although in some cases different co-adapted inversions may become established in different geographic areas and hence contribute to allopatric differentiation. The role of inversion polymorphism in one morabine species, *Keyacris scurra*, has been extensively studied (White 1956, 1957; Lewontin and White 1960; White, Lewontin and Andrew 1963; Turner 1972); but in that case the polymorphisms are broadly the same in the 15-chromosome and the 17-chromosome races of the species, so that they do not seem to be playing any role in speciation. It may well be, however, that some pericentric inversions in other morabine species fall into the category of divisive rearrangements.

The implications of chromosomal fusions and dissociations for the differentiation of populations seem to be quite different to those of pericentric inversions. Although 61 of these are known in the Morabinae, not a single one has been found in a polymorphic state over any geographically extensive area; polymorphism for fusions and dissociations is confined to the zones of overlap, only a few hundred meters wide, between parapatric races or species. Heterozygotes for such rearrangements invariably seem to show, in a proportion of their meiotic divisions, anomalies of synapsis which lead to the production of a certain number of aneuploid gametes and hence to a reduction in fecundity. In these instances the rearrangements are clearly potentially divisive and we can suspect that they may have played a role in speciation.

It seems legitimate to recognize a primitive karyotype in the Morabinae, from which all the others have been derived. This karyotype, with $2n \male = 17$, $2n \female = 18$, occurs in the great majority of the genera, including several which may be morphologically primitive. Of course it is possible that the 17-chromosome karyotype is merely a sort of equilibrium or modal karyotype to which morabines have repeatedly reverted in the course of their phylogeny, and that it is not truly primitive or ancestral, in the sense of having been present in the common ancestor of the morabines. If that is so, the most likely alternative would be a $2n \male = 19$, $2n \female = 20$ karyotype.

The basic and probably primitive karyotype which seems most likely (Figure 1) includes relatively large metacentric 'AB' and 'CD' autosome pairs, six pairs of small autosomes and an X chromosome. The male is XO, the primitive condition in the order Orthoptera. The shapes of the small autosomes in living morabines vary greatly from species to species, so that we cannot guess what they were like in the ancestral morabine; one pair of metacentrics and five pairs of acrocentrics seems to be the most common condition, so that it may be the primitive one. Similarly the X may be either a metacentric or an acrocentric in the living species of morabines, so that cannot tell which form was the ancestral one. One of the small chromosome pairs usually seems to be late-replicating and may also show heteropycnosis at various stages.

Assuming the 17-chromosome karyotype as primitive, we may define the evolutionary changes in karyotype which have occurred subsequently (Table 1 and Figure 2). If the primitive karyotype had $2n \male = 19$, with the A and B

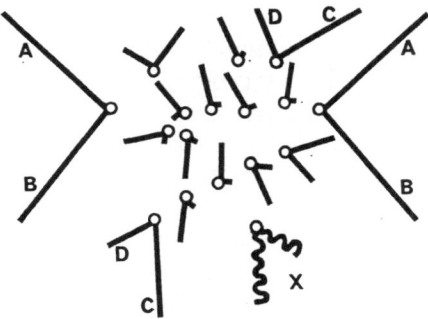

Figure 1. Diagram of the primitive 17-chromosome (♂) morabine karyotype.

elements as separate chromosomes, there would have been about 4 dissociations less and a number of additional fusions in the phylogeny of the morabines.

Table 1. Evolutionary changes of chromosome number in morabine grasshoppers

Fusions			Dissociations	
Centric Type	37		Of the AB chromosome	16
Tandem Type	2		Of the CD chromosome	6
	—			—
	39			22
	—			
X-autosome fusions	11 (1 tandem, 10 centric)			
Y-autosome fusions	6			
Tandem fusion to either X or Y*	1			
Fusions involving sex chromosomes	— 18	18		
	—	—		
Fusions between A and B chromosomes	3			
Fusions between large and small autosomes	8			
Fusions between small autosomes	10			
	—	—		
Fusions between autosomes	21	21		
	—	—		
		39		

* Uncertain which

    With approximately 200 species of Morabinae examined cytologically up till now (about 250 taxa, including chromosomal races) and only 61 fusions and dissociations identified, it is fairly evident that most speciation-events in this group have *not* involved fusions or dissociations. The possibility remains, however, that certain other types of chromosomal rearrangements, including some that are not easily detectable cytologically, may have played a role in some speciation-events similar to that which we believe fusions and dissociations have

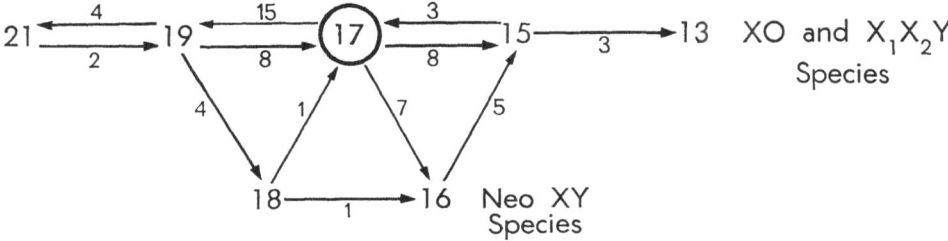

EVOLUTION OF CHROMOSOME NUMBERS
IN THE MORABINAE

39 Fusions ⟶
22 Dissociations ⟵

Figure 2. Diagram showing the changes in chromosome number ($2n \male$) which have occurred in the phylogeny of the Morabinae. The small figures alongside the arrows indicate the number of times a particular change has occurred. Diagonal arrows pointing downward indicate changes from XO ($\male$) to neo-XY; arrows pointing upward indicate changes from XY to X_1X_2Y.

played in the genera which include *viatica* and *virgo*. Two other considerations are relevant here. Some fusions and dissociations are certainly present in the approximately 50 species of Morabines that have not been investigated cytologically. And some fusions and dissociations probably 'cancelled-out' in the past phylogeny of the group, leaving no trace in the surviving species. It is thus probable that between 80 and 100 fusions and dissociations have undergone fixation in the entire phylogeny of the subfamily. This compares with Stone's (1962) estimate of 58 evolutionary chromosome fusions (and no dissociations) in the phylogeny of approximately 2000 species of the genus *Drosophila* known to him at that time. Clearly, there is a much higher probability of changes of this kind undergoing fixation in the Morabinae than in *Drosophila*, a fact which may be related to the difference in population structure, vagility being much lower in the apterous, sedentary Morabinae.

There have been about twice as many fusions as dissociations in the Morabinae. Two explanations for this discrepancy may be suggested. In the first place there are only two chromosome elements in the primitive karyotype (AB and CD) which can undergo successful dissociation, whereas there are usually 5 to 7 that can undergo fusion (depending on how many of the small autosomes are acrocentric). In the second place, it is probable that fusions are inherently more likely to occur than dissociations, because of the nature of the mechanisms involved.

An important difference seems to exist between chromosomal fusions between autosomes and those involving sex chromosomes. The former are always in-principle, reversible—that is to say a fusion may be followed at some later time by a dissociation. X-autosome and Y-autosome fusions, seem, however, to be irreversible. They will be quickly followed by genic changes of a 'degenerative' character in material linked to the Y, leaving the alleles in the X-chromosome or chromosomes hemizygous. Once these changes have occurred the possibility of an evolutionary reversion (from XY to XO or from X_1X_2Y to XY) seems to be out of the question, as far as the Morabinae are concerned. Of the 39 evolutionary fusions that have occurred in morabine grasshoppers 18 involve sex chromosomes (Figure 3). On 11 different occasions in the phylogeny of the subfamily an acrocentric X-chromosome and an acrocentric autosome have undergone fusion so as to convert an originally XO (♂) sex chromosome mechanism into a neo-XY

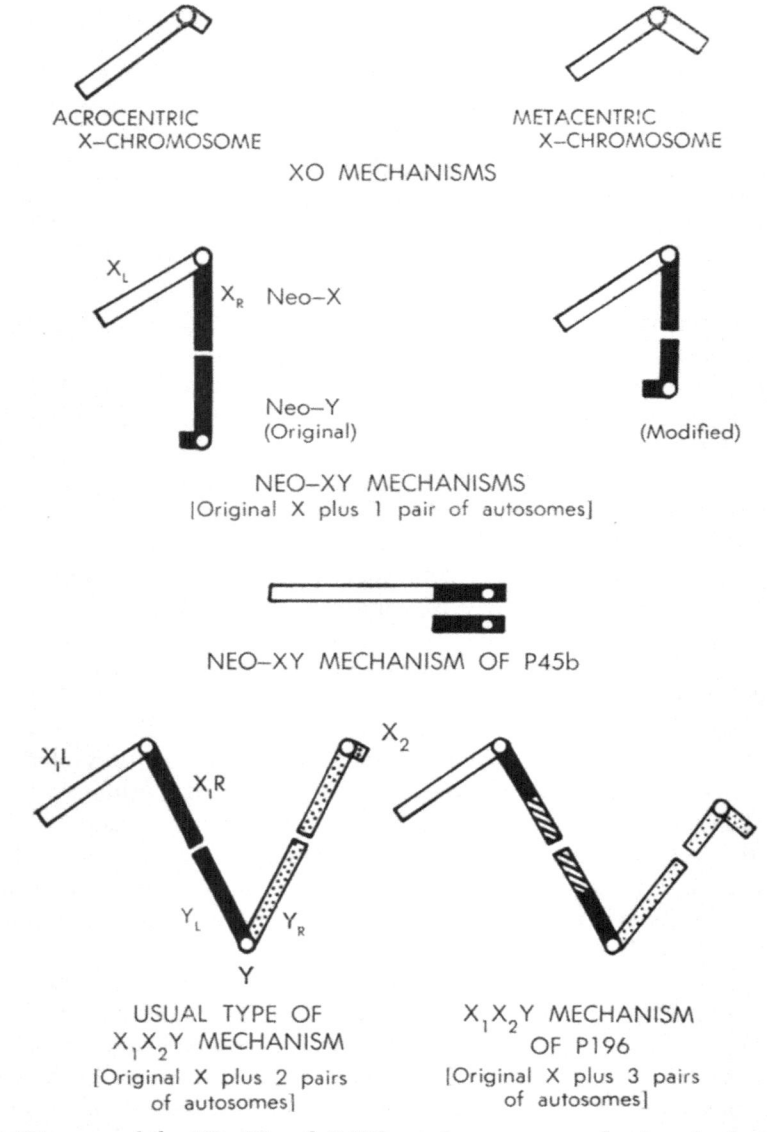

Figure 3. Diagrams of the XO, XY and X_1X_2Y sex chromosome mechanisms in the Morabinae.

one, the neo-Y being simply the unfused homolog of the autosome which underwent the fusion. And in six of these cases a further fusion between the neo-Y and a second autosome has led to the establishment of an X_1X_2Y mechanism, the X_2 chromosome being the homolog of the autosome which fused with the Y (Figure 3). The result of both types of fusions, is, of course, to cause genes that were formerly autosomal to become X-or Y-linked. Originally, that is to say immediately following the establishment of these fusions, chiasma-formation between the fused and the unfused homologous chromosomes is relatively unimpeded, so that sex-linkage would only be partial. But in most instances a process has occurred in evolution which tends to restrict chiasma formation more and more to small distal pairing segments of these chromosome arms, so that the proximal regions come to be genetically isolated from one another because exchanges of material no longer take place between them. Eventually, the neo-Y may be modified in both size and shape, by structural rearrangements so that it is visibly different from the 'right' limb of the X (Figure 3).

A total of fourteen neo-XY races and species of morabine grasshoppers are known to exist at the present time and one more (in the *virgo* group) must have existed formerly. The cytogenetics of nine of these XY forms has been dealt with in White and Cheney (1966, 1972), White, Blackith, Blackith and Cheney (1967), White and Webb (1968) and White Webb and Cheney (1973). There are no large complexes of XY species (the largest being a group of three species in the P52 group), so that neo-XY mechanisms, although they have arisen repeatedly in the morabines, have not in general been long-term evolutionary successes (some of the XY species do, however occupy relatively large areas). X_1X_2Y mechanisms likewise occur mainly in single species or races, some of which are rather restricted geographically. There is, however, one large complex of X_1X_2Y species and races (in the *curvicercus* group, which likewise, includes two neo-XY species); this complex must be regarded as an evolutionary success, since it occurs over a vast area of arid Central Australia, including parts of Queensland, Western Australia and the Northern Territory.

Of the eighteen fusions involving sex chromosomes, sixteen are of the usual 'centric' type. Two, however, are tandem fusions, a type leading at best to semi-sterility in the heterozygote. One of these is the fusion between an acrocentric X and the small late-replicating chromosome 6 in the species P45b of the *viatica* group (Figure 3). The other is a fusion which has occurred in the X_1X_2Y mechanism of species P196 of the *virgo* group (White, Webb and Cheney 1973). In the latter case the sex-chromosome mechanism includes no less than seven chromosomes of the ancestral karyotype.

There seems little likelihood that any of these fusions involving sex chromosomes could ever have existed as balanced polymorphisms and it is thus highly probable that they acted as divisive mechanisms in raciation or speciation. This argument is especially strong in the case of the two tandem fusions, where heterozygotes would have had their fecundity most seriously reduced. But it probably applies also in the case of most of the centric fusions involving the

X-chromosome and small autosomes (eight out of the eleven cases) where the trivalent formed at meiosis in the heterozygote would be a 'lopsided' one.

Two doctrines that have exerted a very considerable influence in evolutionary theory are that complete geographic isolation between two populations (allopatry) must precede the acquisition of genetic isolating mechanisms between them and that so-called premating isolating mechanisms (mainly but not exclusively ethological) have played a primary role in speciation since only they (and not the post-mating isolating factors) are stated to be capable of being strengthened by natural selection.

Both these doctrines were challenged, as far as the morabine grasshoppers, and other organisms having a similar type of population structure, are concerned, by the 'stasipatric' model of speciation put forward by White, Blackith, Blackith and Cheney (1967) and White (1968). The essential elements of this model, originally derived from the *viatica* species group of the Morabinae, have been explained by Key (1973) although he does not use the term 'stasipatric' and does not entirely accept all elements in the model. There are 12 taxa in the *viatica* group, whose distribution and karyotypes are shown in Figure 4. The

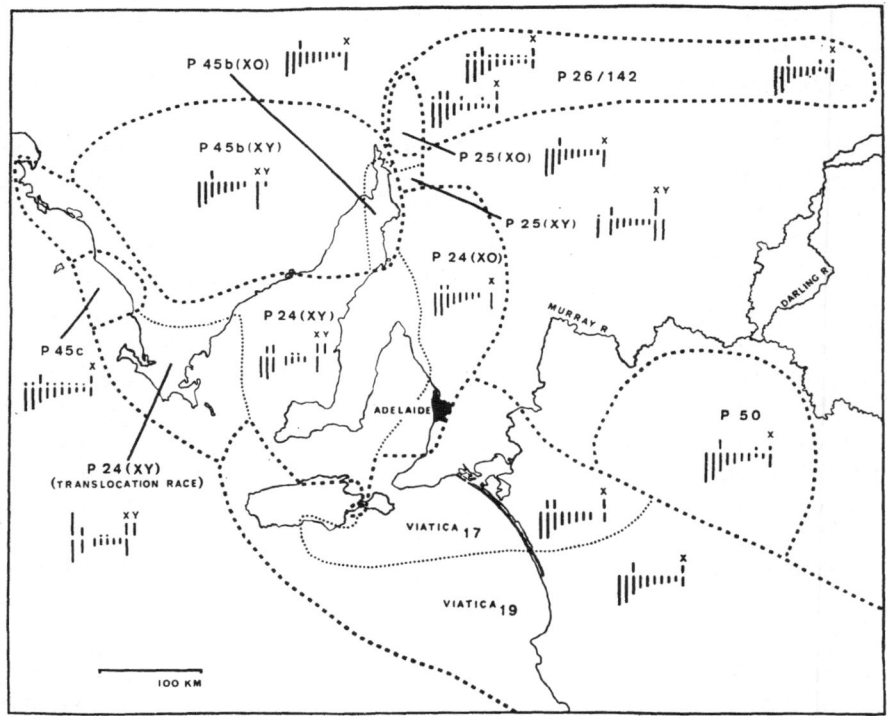

Figure 4. Map showing the distribution of the taxa in the *viatica* group, with ideograms of their karyotypes. Forms here regarded as species separated by dashed lines, taxa regarded as chromosomal races separated by dotted lines. The 'inland species' are *P*50 and *P*26/142 (with several different karyotypes). The 'coastal species' recognized here are *viatica*, *P*24, *P*25, *P*45b and *P*45c.

karyotypes of the 'coastal' taxa differ mainly by major rearrangements (fusions and a translocation) in the ways discussed by White, Blackith, Blackith and Cheney (1967). P45c differs from *viatica* in that the short arms of all the acrocentric chromosomes are significantly longer; since this material is late-replicating (Webb 1973) it is likely that in the evolution of P45c the length of the short arms has been increased by multiple duplications of material. In the 'inland' species P26/142, on the other hand, where the short arms of some or all of the acrocentrics have undergone an increase in length they are not late-replicating (Webb 1973) so that it is likely that pericentric inversions rather than duplications have occurred.

Stasipatric speciation involves the origin of chromosomal rearrangements giving adaptively superior homozygotes but inferior heterozygotes within the population of a species. Although most such rearrangements would undoubtedly be eliminated by natural selection, it is believed that on rare occasions one of them may spread geographically through the territory occupied by the species and initiate a divergence between the population carrying it and the surrounding population. Since the stasipatric model was first proposed, it has been plausibly suggested that it may also be applicable in the case of some stick insects (Craddock 1973) and in some small mammals, especially fossorial ones (White 1973).

Our general conclusion has always been that ethological isolation is minimal or even non-existent as far as closely related morabine species are concerned. There is no stridulation and apparently no complex courtship behaviour. Nevertheless, even though copulations between different taxa may be quite prolonged, it is possible that failure of sperm transfer (or at any rate effective sperm transfer) may indicate that such copulations are not ethologically quite normal. In the case of the parthenogenetic species *virgo*, we have found that copulation can occur quite readily in the laboratory, with males of several different species from Western Australia, and that in some instances triploid hybrids can be obtained.

In the published discussions of speciation mechanisms there seems to have been an implicit divergence between those who believe that all the individual steps in the process must have been almost imperceptably small, i.e. due to mutations having only slight effects, and those who take the view that, at least in some instances, major sudden changes (of which chromosomal rearrangements could be an important category) may play a special role. Those workers who study the ethological factors in speciation have not explicitly denied that karyotype evolution may be involved in speciation but their total silence on the subject seems to imply that it can be left out of account. However, it is evident that in a group like the Morabinae the major chromosomal rearrangements, and especially those whose heterozygotes are adaptively inferior, have somehow to be fitted into the overall evolutionary picture. And it is extremely unlikely that they can be regarded as simply equivalent to the other mutations, that is, those of single genes. Thus even if the stasipatric hypothesis is not accepted in its entirety, a fully complete model of speciation will have to include major chromosomal rearrangements in the case of all groups where the species show differences in karyotype.

Key (1973) has outlined the general nature of the role which we believe chromosomal rearrangements have played in the speciation of the *viatica* complex. He differs from the writer on two matters, both perhaps of minor importance to the main argument. In the first place he attributes much greater importance to the possibility that chromosomal rearrangements may establish themselves on the periphery of the species-distribution.

Now only a minute fraction of the individuals or local colonies of a species are strictly peripheral, geographically. The vast majority exist within the body of the species-population, in non peripheral situations, although because of the patchy discontinuous or semi-continuous distribution pattern of morabine populations, which is likely to have existed even before the coming of European man to Australia (and even before the arrival of Aboriginal man), many individuals and local colonies may have existed at or close to 'internal peripheries'. It is hence statistically in the highest degree improbable that the rare rearrangements which are potentially isolation-generating would occur precisely on the periphery of a species distribution. However, it has long ago been pointed out that natural selection is a mechanism which produces situations that are highly improbable on a random basis, and it could be shown that selection is extremely severe against newly arisen major rearrangements in the centre of the species distribution, we would have to consider Key's argument more seriously than seems warranted at present.

However, the question whether the initial origin of the rearrangement is peripheral or not does not seem to be of prime importance. If it spreads through the pre-existing range of the species we must regard the process as a stasipatric one, rather than allopatric. On the other hand, a chromosomal rearrangement which spreads out into previously unoccupied territory may be regarded as playing a role in an allopatric process (Figure 5). It is quite possible, and even probable, that both processes have occurred in the Morabinae.

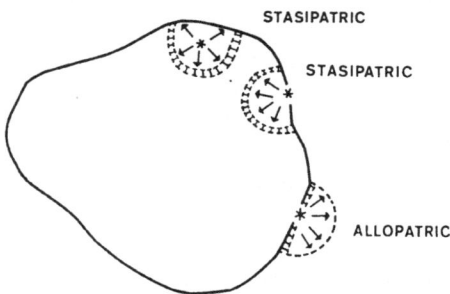

Figure 5. Diagram showing stasipatric and allopatric spread of chromosomal rearrangements from their point of origin.

In the second place, Key regards the taxa of the coastal complex in the *viatica* group as a single species with ten chromosomal races, whereas the writer believes that we should recognize five or six species. This is not merely a question of semantics. To recognize a species within which nine different karyotypes exist and which is subdivided into taxa separated by very major reproductive barriers, some apparently of a physiological nature, would imply the recognition in the Morabinae of species in no way comparable to the species of *Drosophila*, *Anopheles* or *Chironomus* (to pick insect examples that have been thoroughly studied from an evolutionary standpoint).

Key has made reference to two species concepts, the one based on reproductive isolation (due to Mayr, 1942) and the one based on irrevocability of divergence of two forms, due to Simpson (1961) and Bigelow (1965). One might perhaps suggest a third concept, (not as an alternative to the other two, but supplementary to them) namely that of the species as a unique co-adapted genetic system. None of these concepts is particularly easy to apply in the case of the Morabinae. But the third one may help us in the case of a form like P45c of the *viatica* group where multiple duplications have built up a quantity of late-replicating DNA in the short limbs of all of the acrocentric chromosomes. The total amount of duplicated DNA and the number of short limbs involved is higher in P45c than in the other members of the *viatica* group; in particular it is higher than in P24 and P45b which occupy adjacent or nearby areas of the total distribution.

Although the stasipatric model seems a useful one for understanding some speciation events in the Morabinae, it is not necessarily suggested that it is fully applicable in every case. And it is unlikely to apply at all in organisms such as *Drosophila,* whose vagility is greater by one or two orders of magnitude than that of the morabine grasshoppers. It may also be inapplicable in groups where ethological isolation, based on sound, scent or courtship behaviour, plays a major role in speciation. But the time seems to have come for us to realise that speciation has followed different patterns in different groups, so that a comparative study of speciation mechanisms will recognize a number of different models.

REFERENCES

Bigelow, R. S. 1965. Hybrid zones and reproductive isolation. Evolution 19: 449-458.
Craddock, E. M. 1973. Article in this volume.
Key, K. H. L. 1973. Article in this volume.
Lewontin, R. C. and White, M. J. D. 1960. Interaction between inversion-polymorphisms of two chromosome pairs in the grasshopper *Moraba scurra*. Evolution 14: 116-129.
Mayr, E. 1942. Systematics and the Origin of Species. Columbia Univ. Press.
Simpson, G. G. 1961. Principles of Animal Taxonomy. Columbia Univ. Press.
Stone, W. S. 1962. The dominance of natural selection an the reality of superspecies (species groups) in the evolution of *Drosophila*. Univ. Texas Publ. 6205: 507-538.
Turner, J. R. G. 1972. Selection and stability in the complex polymorphism of *Moraba scurra*. Evolution 26: 334-343.
Webb, G. C. 1973. Unpublished Ph.D. thesis (University of Melbourne).
White, M. J. D. 1956. Adaptive chromosomal polymorphism in an Australian grasshopper. Evolution 10: 298-313.
White, M. J. D. 1957. Cytogenetics of the grasshopper *Moraba scurra*. II. Heterotic systems and their interaction. Austral. J. Zool. 5: 305-337.
White, M. J. D. 1968. Models of speciation. Science 159: 1065-1070.

White, M. J. D. 1973. Chromosomal rearrangements in mammalian population polymorphism and speciation. In: Cytotaxonomy and Vertebrate Evolution (ed. B. Chiarelli and E. Capanna). London: Academic Press.

White, M. J. D., Blackith, R. E., Blackith, R. M. and Cheney, J. 1967. Cytogenetics of the *viatica* group of morabine grasshoppers. I. The 'coastal' species. Austral. J. Zool. **15**: 263-302.

White, M. J. D. and Cheney, J. 1966. Cytogenetics of the *cultrata* group of Morabine grasshoppers. I. A group of species with XY and X_1X_2Y sex chromosome mechanisms. Austral. J. Zool. **14**: 821-834.

White, M. J. D. and Cheney, J. 1972. Cytogenetics of a group of morabine grasshoppers with XY and X_1X_2Y males. Chromosomes Today (ed. C. D. Darlington and K. R. Lewis) 3: 177-196: London: Longman group.

White, M. J. D., Lewontin, R. C. and Andrew, L. E. 1963. Cytogenetics of the grasshopper *Moraba scurra*. VII. Geographic variation of adaptive properties of inversions. Evolution **17**: 147-162.

White, M. J. D. and Webb, G. C. 1968. Origin and evolution of parthenogenetic reproduction in the grasshopper *Moraba virgo* (Eumastacidae: Morabinae). Austral. J. Zool. **16**: 647-671.

White, M. J. D., Webb, G. C. and Cheney, J. 1973. Cytogenetics of the parthenogenetic grasshopper *Moraba virgo* and its bisexual relatives. I. A new species of the *virgo* group with a unique sex chromosome mechanism. Chromosoma **40**: 199-212.

II

EVOLUTION IN THE HAWAIIAN DROSOPHILIDAE

Introduction and background information[1]

D. ELMO HARDY

Senior Professor of Entomology, University of Hawaii, Honolulu, Hawaii 96822

Since 1963 a team research, multi-disciplinary project has been underway attempting to gain as much information as possible concerning evolutionary processes under insular conditions: why some groups have speciated so profusely in the Hawaiian Islands; why and how such an array and diversity of morphological and biological characters have arisen; the time required for speciation; and ultimately the origin of species. This has been a cooperative project of the University of Hawaii Departments of Entomology and Genetics and the Genetics Foundation, University of Texas, and has involved many senior investigators, technicians, and assistants. To date approximately 25 senior scientists from various institutions and about 150 graduate and undergraduate research assistants have taken part in this study. The stock center for the Hawaiian species has been established and is being maintained at the University of Texas. The species now in culture are available for experimental use by qualified researchers.

The results of these studies have been published in 61 papers, mostly in the University of Texas Publications, Studies in Genetics, and a detailed summary of accomplishments was published in a special volume of Evolutionary Biology honoring Professor Theodosius Dobzhansky (Carson, *et al.*, 1970).

The dipterous family Drosophilidae is remarkably developed in Hawaii and represents one of the most striking examples of rapid, adaptive radiation known in the animal kingdom (Hardy, 1965). To the present time we have over 500 named species and now estimate that the total fauna may number 750-800 species. As reported by Wheeler and Hamilton (1972), 279 species of endemic Hawaiian *Drosophila* have been described constituting 'an amazing 53.3 per cent of all the names proposed since 1959!' Actually 18 additional species should be added to this list which were placed in other combinations but which we now feel should be under *Drosophila,* so that the total number recorded in the past 13 years is 297 species. The publication of volume 12 (Hardy, 1965) of the Insects of Hawaii which dealt with the family Drosophilidae,

1 Published with the approval of the Director of the Hawaii Agricultural Experimental Station as Journal Series No. 1577.

recording 400 species from these islands, served as the background upon which the project on Evolution and Genetics of the Hawaiian Drosophilidae has been developed. Fifty-six species of *Drosophila* were described before 1959 bringing the total number of described species in this genus to 353. Adding the 132 species of scaptomyzoids the number of described Drosophilidae is now 485 and with the 22 immigrant species the fauna now totals 507 species. As cited by Wheeler and Hamilton (*loc. cit.*) a total of 1,254 species of *Drosphila* have been described for the entire world, over one-fourth of these are native to the Hawaiian Islands.

The Hawaiian drosophilid fauna was discovered by R.C.L. Perkins while doing the field work for the Fauna Hawaiiensis in the 1890s. In his (1913) 'Introduction' he said that the genus *Drosophila* 'is represented by an assemblage of species, exhibiting great diversity in structure and appearance. . . . At present these insects, many of which are obscure and minute forms, have been very imperfectly collected. To make an approximately complete collection and thorough study of the Hawaiian species would require the devotion of many years of special work. Not less than 250 species must exist in the islands, and double that number may very probably occur.' It is interesting to note that Perkins collected 47 species. These were described by Grimshaw (1901, 1902) and by Perkins (1910) and all are *Drosophila*. Grimshaw described *Idiomyia* and *Hypenomyia* as new genera but these have since been synonymized with *Drosophila*. A few additional new species were added to the family by Bryan (1934, 1938), Malloch (1938), Wirth (1952), and Hackman (1959, 1962). The number of described species in the family from Hawaii previous to my own studies was 73.

A little known story in the history of the Hawaiian Drosophilidae is the preliminary study made by G. B. Mainland who was trained at the University of Texas under J. T. Patterson and W. S. Stone and who came to Hawaii about 1946 with the intent of beginning studies on the genetics of the native *Drosophila*. He began this work and apparently was partially successful in getting two species, *pilimana* Grimshaw and *crucigera* Grimshaw into laboratory cultures. He used standard *Drosophila* media and probably was not able to maintain the flies for more than one or two generations. The only information which came out of his preliminary study was in a paper by Hederick and Burke (1950) entitled 'Yeasts from Hawaiian Fruit flies: Their Identification and Ability to produce Riboflavin.' Dr. Mainland sent them mixed microbe cultures of *pilimana*, *crucigera* and a 'small, brown *Drosophila*' on Difco AC agar streaks and stabs from feces, crop content and the immediate natural substrate. Seventeen isolates of yeast were obtained which were resolved into three genera and six species. This was a very important background study. According to E. C. Zimmerman (1958), Professor Richard Goldschmidt was interested in doing genetics studies of the Hawaiian species in the early part of this century and his plan for coming to Hawaii to begin such a study was thwarted by the outbreak of the first world war. He later planned to send one of his students to Hawaii to study the *Drosophila*, and had mentioned the possibility to Curt Stern, but again this was prevented by the outbreak of the second world war. Zimmerman (*loc. cit*) was well aware of the unusual drosophilid fauna and made a strong plea for evolutionists and geneticists to do work on this remarkable fauna before it is too late. He wrote 'as a student of some years experience in island faunas, I wish to urge those who are equipped to do specialize research on *Drosophila* to look to the Hawaiian drosophilid fauna as a veritable mine of possibilities for exciting and profitable study.'

The urgency for studies of the endemic biota is obvious. We have evidently lost more of our native species than any area of comparable size over the world, and have more species on the rare and endangered list of world animals. Dr. A. J. Berger (personal communication) has stated that of the Hawaiian birds more species are on the rare and endangered list than are listed from all of the continents combined. The island ecosystems are extremely fragile and the changes being brought on by man and introduced animals and plants are alarming. Urbanization, destruction of the forests and native plants, onslaughts of introduced herbivores, predators and parasites are obviously eliminating the native species. Over the past 25 years an average of 16 new immigrant insects have become established in the islands each year. Most of these remain in the lowlands but some have spread widely and have become established in native habitats. Almost without exception when immigrant species move into niches occupied by endemics the latter are unable to survive the competition pressures. The endemic biota is almost a total loss in the lowlands of the islands. We have no way of knowing what this biota might have been. The native plants and animals, almost without exception are now found only at higher elevations, 2000 feet or over, or deep in the valleys where some native plants are still holding on. It is evident that the area now occupied by native species is much less than half what it once was.

My first introduction to the Hawaiian drosophilids came 25 years ago shortly after first coming to Hawaii the fall of 1948. While on a field trip with Dr. Mainland in the Waianae mountains of Oahu, I saw my first 'Idiomyia' species sitting on a tree fern fronds. My first reaction was one of disbelief and from my experience in dealing with Diptera of other parts of the world I couldn't conceive of it being a *Drosophila*. Subsequently, while doing the survey work for the Diptera of Hawaii volumes, especially the volume dealing with the Drosophilidae, I became more and more impressed that here indeed was a remarkable group of organisms which would be ideal for evolutionary studies and which should be exploited. Over the years I began telling the story of these unusual flies to evolutionists, geneticists, and anyone who would listen, wherever I went. It wasn't until 1961 that I found 'receptive ears' when I visited Dr. W. S. Stone, University of Texas, Genetics Foundation. After hearing 15 minutes of my story Dr. Stone said, 'Let's do something about it.' We immediately made plans for a cooperative research project between the University of Hawaii and the University of Texas, Genetics Foundation and thanks to Dr. Stone's organizing genius we were able to get financial support from the National Institutes of Health and the National Science Foundation and the project got underway in April 1963. By this time a taxonomic arrangement of the family was completed and this provided the necessary background information to get the project started.

The pioneer group of investigators who came to Hawaii consisted of Drs. W. S. Stone, H. L. Carson, Frances E. Clayton, W. B. Heed, H. T. Spieth, H. D. Stalker, L. H. Throckmorton and M. R. Wheeler. All of these people had many years of experience working with *Drosophila* both in the field and laboratory and with their expertise the ground work for the project was laid the first summer. Nevertheless the initial period was largely one of orientation and 're-education of the experts'. For someone used to working with *Drosophila* over the rest of the world the first encounter with most of our Hawaiian species is almost always one of mild shock or disbelief. Newcomers to the scene have to learn that one doesn't collect Hawaiian species in the usual situations or with the techniques normally used. They also must learn that these flies are different from droso-philids from other parts of the world in almost every way which one cares to analyze them. Many are the giants of the family, some have a wing-spread of ca. 18-20mm. On the other end of the size scale some of the endemic species are among the smallest known drosophilids (ca. 1.5 mm). The flies show great diversity of morphological, behavioral, and ecological characteristics. A majority of species are highly ornate, have a great variety of patterns or maculations in the wings and exhibits unusual sexual dimorphisms, with all sorts of strange developments of the male labella, legs, and other body parts which are not found in other drosophilids. As our studies have proceeded it has become increasingly more apparent that for the most part the strange morphological characters developed in the Hawaiian species are of little, and often no value in determining relationships. It is quite impossible, and com-pletely unreliable to base phylogenies upon morphology alone and evolutionary trends have to be determined by pooling knowledge gained by cytological studies, hybridization, behavior, ecology, isozyme variations, and morphology.

The Hawaiian species have radiated into a great variety of habitats (Heed 1968, Carson, *et al.*, 1970) and for the most part their food and substrate pre-ferences differ from those of species from other areas. They breed in an assortment of situations such as decaying leaves, stems, bark, flowers, and fruits of native plants; fleshy fungi; morning glory flowers; slime fluxes; in soil wet by dripping fluxes; frass in stem mines of other insects; plant hairs of *Cyrtandra*; and as predators on thomisid spider eggs.

The first job to be done was to learn how to get the native species into laboratory cultures. Previous to this time except for the brief attempts of Dr. Mainland no one had been successful in getting species into artificial media and no genetic information had ever been obtained. Unfortunately Dr. Mainland left no information behind, it would no doubt have given us some sort of a starting point if we had had access to information he gained from his experimenta-tion. Attempts with standard *Drosophila* media were unsuccessful and a great deal of field and laboratory work had to be done before we were able to learn the nutritional and substrate requirements. Our first major breakthrough came when Drs. M. R. Wheeler and F. E. Clayton (1965) developed a food medium and acquired the necessary laboratory know-how to begin raising some of the picture-winged species. This period was one of considerable frustration and much trial and error experimentation and an element of serendipity was frequently involved in the discoveries made. An early baffling problem was to find the substrate requirements of the females for oviposition. Drs. Wheeler and Clayton spent weeks trying everything they could dream up to get the females to deposit their eggs. The gravid specimens brought in from the field would be very happy in food vials but could not be enticed to oviposit. Finally out of desperation a few moistened bits of Kellogg's Special K breakfast food were thrown into one of the cages. The females went immediately to these bits of cereal and began stuffing them with eggs. As soon as knowledge of the natural substrates became available through the ecological studies of Dr. Heed (1968) and later Steve Montgomery (in press) there were no further problems getting most of the picture-winged species to oviposit.

The next frustration in developing the laboratory techniques was in figuring out what kind of a medium the mature larvae needed for pupation. During the trial-and-error days the larvae would mature in the food medium and then spend all their time wandering around the sides of the vial trying to escape. Finally the food vials were placed in a jar of moist, sterile beach sand and the larvae allowed to crawl out of the vials. They took immediately to the sand and the problem was solved; they completed their development. Improvements in food media and techniques for handling have continued to be made by the laboratory workers handling the flies, and especially by Miss Kathleen Resch at the University of Texas whose 'tender, loving care' of the Hawaiian *Drosophila* stocks has resulted in outstanding achievements in getting a large number of species into cultures. Almost complete success has now been had with all of the picture-winged species and Miss Resch and her assistants are now achieving good results with 'modified mouth parts' and some other species groups with which she had been working. Approximately 140 species have now been reared in laboratory media.

In spite of the fact that outstanding successes have been achieved and a great deal of information is now at hand concerning the nutritional and breeding requirements of native species we still are able to handle only about one-fourth or one-fifth of the total fauna in the laboratory, so much still remains to be learned. One important discovery has been that extracts of *Clermontia* leaves and stems have fungicidal properties and by using this in our media we are able to reduce fungal growth to manageable proportions.

The original survey upon which volume 12 was based was conducted entirely by use of insect nets, mostly beating the flies out of vegetation by using a heavy beating bag; this was supplemented by some trapping, using fermenting banana baits. The early field work of the research team used the standard *Drosophila* net (Wheeler net) and baiting techniques. During the first season or so all sorts of rotting fruit and vegetable concoctions were tried, with poor success. The first exciting field accomplishment was when Dr. Heed discovered *Drosophila* breeding in rotting leaves of *Cheirodendron* at Paliku, Haleakala Crater, Maui. This was soon followed by the discovery that many species also live in rotting bark and stems of *Clermontia* and other lobeliads, also leaves and other parts of a variety of native plants, fleshy fungi and an assortment of other habitats. Then after considerable experimentation a very successful bait was developed using strained banana baby food inoculated with yeast. The field workers soon learned the types of habitats frequented by the flies, aided by the information gained from Dr. Heed's ecological studies and Dr. Spieth's courtship and mating behavior studies, and by applying the fermenting banana bait on trunks of trees, under limbs and other spots in deep shade they could quickly attract flies. The collecting procedure has progressed to the point where it is done largely by a visual technique. The field personnel now rarely bother to carry nets but do their collecting by setting up trapping lines in favorable spots in the forest and sampling these at frequent intervals by watching for the flies sitting on trunks, stems, or leaves, often feeding on or sitting near the bait. When a fly is spotted a 25 by 95 mm vial is quickly placed over it and the fly pops into the bottle. With practice, a sharp eye, and a degree of skill one is able to capture a high percentage of individuals. This technique has turned up a large number of species which have never been collected by other methods, also some species which have previously been considered rare were found common once their host plant and habitats were known. The flies are then transferred to food vials, containing blotting paper soaked in a sugar-agar solution and placed in cool boxes for transportation to the laboratory. Prior to the initiation of this project attempts to bring live material from other islands into the laboratory failed because we did not utilize insulated boxes. We now have a high degree of success bringing live adults and immature states in as long as they are kept in a cool, moist environment. Likewise the rearing laboratory is maintained at ca. 64 degrees F. (18-20°C) and the humidity 70°+ degrees to simulate temperature and humidity found in the natural environments.

As pointed out above a major challenge remains to solve the nutritional and ecological problems presently preventing us from investigating a major portion of the Hawaiian Drosophilidae. Progress is rapidly being made toward this end and successes are now being had at Texas rearing leaf-breeding species. This should make it possible to apply the same research techniques used so successfully for the picture-winged species to other major species groups. Until this knowledge is available it will be impossible to do more than speculate on many of the phyletic lines from the *Scaptomyza* through the *Drosophila*.

Due probably to my background as a general collector and my introducing the investigators to the highly developed rain forest fauna, for many years we assumed that the native species were predominantly breeders in the wetter areas of the islands. Areas with annual rainfalls of 100-300 plus inches per years were considered our best collecting spots. It became a maxim that the collectors expected to get wet whenever they had a successful day in the field, that the flies were happiest when it was raining. In fact in the rain forest the collecting is best during a light drizzle, the flies show preference for absolute humidity and in some areas activity decreases sharply when the humidity drops below 90 per cent. It came as somewhat a surprise when we investigated the dry south slope of Haleakala on Maui. In the Auwahi area, 3000-4000 ft. elevation, where a remnant of native dry forest remains, we found a rich fauna of *Drosophila* and in the past three years a graduate student, Mr. Steve Montgomery, has made an intensive survey of the *Drosophila* living in the dry land forests throughout the islands. This has opened up an entirely new area for investigation and has greatly increased our knowledge of the breeding habits of these flies. Mr. Montgomery (in press) has been successful in obtaining host plant information for 76 species of native picture-winged *Drosophila,* he has reared over 2,000 specimens of picture-wings from 59 host plant species belonging in 33 plant families. Some species are general polyphagous feeders while others are very host specific. Mr. Montgomery has turned up many new species and a great many host and substrate associations for species which were previously not known. His work is of major importance in our understanding of the Hawaiian fauna.

The native drosophilids show a high degree of single island endemism. It is probable that 98 per cent of the species are restricted to a single island and it now appears likely that speciation has occurred in isolated areas (volcanic masses, mountains) on various islands. With few exceptions whenever a 'species' which has been reported to be on two or more islands is investigated in detail; it is found to represent a complex of closely related species. It is apparent that various isolated volcanic masses (mountains) on different islands and also 'kipukas' (islands of vegetation in lava fields) create effective barriers against gene flow between adjacent populations and depending upon spatial isolation and geologic time involved these provide excellent 'laboratories' for studying all degrees of speciation. We have been treating the complex of closely situated islands, Maui, Molokai, and Lanai, as one biological island since these have been joined and separated two or more times in rather recent geological history. It has been evident that these islands have many species and groups in common. In making close analysis of these 'species', however, it has become apparent that many of them have become full species and we have been able to demonstrate through crossmating experiments that a number which had been considered one species found on both Maui and Molokai are actually two. These are morphologically indistinguishable (sibling) but show reproductive isolation.

As has been demonstrated by Carson using polytene chromosome mapping techniques (Carson, *et al.* 1970) a main center for adaptative radiation has been the Maui complex of islands. In certain significant cases the ancestoral founder species came to Maui direct from Kauai, a flowering of species occurred on this island with founders then radiating out to Hawaii, Oahu, and in at least one case back to Kauai. Of the drosophiloids which have been described to date approximately 41 per cent are from the Maui complex; 25 per cent are from the island of Hawaii; 22 per cent from Oahu; and only 12 per cent from Kauai. This is not an exact indication of the percentages of species on the various islands and does reflect somewhat the areas which have received greatest attention to date. It is

obvious, however, that the greatest amount of speciation has occurred on the
Maui complex and by comparison the fauna of Kauai is rather sparse. *Pheidole*
ants have made extensive inroads on Kauai, they occur over most of the island
to about 4000 feet and have no doubt had a drastic effect upon the native fauna.
It is dubious that most drosophilids, especially the leaf-breeding species, could
survive the biotic pressures exerted by such effective predators. We have 42
introduced species of ants in the Hawaiian islands and many biologists feel that
these animals have probably been responsible for elimination of many species
of endemics.

It has been commonly expressed that a probable reason for the profusion
of species which have evolved in many groups of Hawaiian animals is that the
successful ancestral immigrants found ample unoccupied ecological niches and
met with little or no predators, parasites, or diseases. This is definitely not the
case in the Drosophilidae. A large fauna of about 100 species of predaceous
muscid flies, *Lispocephala,* live in close association with drosophilids in both
the larval and adult stages. The flies are also preyed upon by endemic native
birds and crabronid wasps and the leaf-breeding forms no doubt serve as prey
for Dolichopodidae. About 250-300 native species of the latter occur and are
abundant on the ground and vegetation in the rain forest. Also Eucoilinae wasps
are frequently encountered in the drosophilid habitats but their effects have not
yet been studied. Spieth, (Carson, *et al.* 1970) has indicated that not only
courtship behavior but all other aspects of the drosophiloids behavior indicate
high adaptation to the avoidance of predators, and this has been an important
factor in their evolution.

Internal morphological studies (Throckmorton 1966) have demonstrated
that the reproductive system is remarkably stable and much more reliable for
determining gross relationships than are external characters. The result of these
studies augmented by egg structure, female ovipositors, behavior, and male
genitalia studies, shows that the two major groups, drosophiloids and scaptomyzoids, intergrade in Hawaii and indicates that the total drosophilid fauna
of these islands could have originated from one ancestral species. This leads to
a rather startling implication that the genus *Scaptomyza* Hardy could have arisen
in Hawaii. Adding even more complexity to the story, however, we now know
87 species of *Scaptomyza* (*Trogloscaptomyza*) Frey from the entire world; 86
occur only in Hawaii and the type of the subgenus, *S. brevilamellata* (Frey) was
described from the island of Tristan da Cunha in the middle of the Atlantic
Ocean.

The volume of Insects of Hawaii dealing with the Drosophilidae was a
phenetic taxonomic treatment with no attempt at phyletic arrangement except
as species segregated in the keys according to morphological characters. With the
evidence now at hand from the many different types of research findings it is
evident that the species of Hawaiian Drosophilidae must be arranged in only
three genera. Previously nine 'genera' have been proposed, seven of these under
drosophiloids, are all *Drosophila,* 6 being direct synonyms and only *Antopocerus*
Hardy probably representing a valid subgenus. Under the scaptomyzoids the

spider egg predators are being retained as being a distinct genus, *Titanochaeta* Knab.

For convenience the majority of the *Drosophila* have been arranged in groupings based upon prominent morphological characters, such as: picture-wings, modified mouthparts, ciliated tarsi, bristle tarsi, fork tarsi, spoon tarsi, and white-tipped scutellum (Heed 1968). To date the major research efforts have been with the picture wings (ca. 120 species) since these could be handled in the laboratory. Such detailed information as we now have on the picture-winged species has rarely been obtained for such a group of animals and the combined efforts of our research team will result in a classical example of the proper approach to systematics and evolution. We now need to apply the same approach to other major groups of Drosophila and eventually to the *Scaptomyza*. We still have only a preliminary knowledge of the latter and it is probable that 250-300 native species occur in these islands.

ACKNOWLEDGEMENT

The evolutionary studies of the Hawaiian Drosophilidae have been made possible by the financial support of the National Institutes of Health grant GM 10640 and National Science Foundation grant GB 29288.

REFERENCES

Bryan, E. H. 1934. A review of the Hawaiian Diptera, with descriptions of new species. Proc. Haw. Ent. Soc. 8 (3): 434-440 and 456-457.

Bryan, E. H. 1938. Key to the Hawaiian Drosophilidae and descriptions of new species. Proc. Haw. Ent. Soc. 10 (1): 25-42.

Carson, H. L., D. E. Hardy, H. T. Spieth and W. S. Stone. 1970. The Evolutionary Biology of the Hawaiian Drosophilidae. Essays in Evolution and Genetics in honor of Theodosius Dobzhansky. A suppl. to Evolutionary Biology. Appleton-Century-Crofts, New York. pp. 437-543.

Grimshaw, P. H. 1901. Fauna Hawaiiensis 3 (1): 51-73.

Grimshaw, P. H. 1902. Fauna Hawaiiensis 3 (2): 86.

Hackman, W. 1959. On the genus Scaptomyza Hardy (Dipt. Drosophilidae) with descriptions of new species from various parts of the world. Acta Zool. Fennica 97: 3-73.

Hackman, W. 1962. On Hawaiian Scaptomyza species (Dipt. Drosophilidae). Notulae Ent. 42: 33-42.

Hardy, D. E. 1965. Insects of Hawaii, vol. 12. Diptera: Cyclorrhapha II, Series Schizophora, Section Acalypterae I. Family Drosophilidae. Univ. of Hawaii Press, Honolulu. 814 pp.

Hederick, L. R. and Burke, G. C. 1950. Yeasts from Hawaiian fruit flies: their identification and ability to produce riboflavin. J. Bact. 59: 481-484.

Heed, W. B. 1968. Ecology of the Hawaiian Drosophilidae. Univ. of Texas Publ. 6818: 387-419.

Malloch, J. R. 1938. Two genera of Hawaiian Drosophilidae (Diptera). Proc. Haw. Ent. Soc. 10 (1): 53-55.

Perkins, R. C. L. 1910. Fauna Hawaiiensis 2 (6) Suppl. to Diptera, pp. 697-700.

Perkins, R. C. L. 1913. Fauna Hawaiiensis, Intro. 1 (6): CLXXX-CLXXXIX.

Wheeler, M. R. and Clayton, F. E. 1965. A new Drosophila culture technique. Drosophila Inform. Serv. 40: 98.

Wheeler, M. R. and Hamilton, N. 1972. Catalog of Drosophila species names, 1959-1971. Univ. of Texas Publ. 7213: 257-268.

Wirth, W. W. 1952. Two new spider egg predators from the Hawaiian Islands (Diptera: Drosophilidae). Proc. Haw. Ent. Soc. 14 (3): 415-417.

Zimmerman, E. C. 1958. 300 species of Drosophila in Hawaii?—A challenge to Geneticists and Evolutionists. Evolution 12: 557-558.

LIST OF PAPERS DEALING WITH HAWAIIAN
DROSOPHILIDAE PUBLISHED SINCE 1970 REVIEW

Carson, H. L. 1970. Chromosomal tracers of founder events. Biotropica **2** (1): 3-6.

Carson, H. L. 1970. Chromosome Tracers of the Origin of Species. Science **168**: 1414-1418.

Carson, H. L. 1971. Speciation and the Founder Principle. Univ. of Missouri, Stadler Symposia **3**: 51-70.

Carson, H. L. 1971. Polytene chromosome relationships in Hawaiian species of Drosophila V. Additions to the chromosomal phylogeny of the picture-winged species. Univ. of Texas Publ. **7103**: 183-191.

Carson, H. L. 1971. The ecology of Drosophila breeding sites. Univ. of Hawaii, Harold L. Lyon Arboretum Lecture **2**: 1-27.

Clarke, B. 1970. Review in Science **169** (3951): 1192.

Clayton, F. E. 1971. Additional karyotypes of Hawaiian Drosophilidae. Univ. of Texas Publ. **7103**: 171-181.

Clayton, F. E., Carson, H. L. and Sato, J. E. 1972. Polytene chromosome relationships in Hawaiian species of Drosophila. VI. Supplementary data on metaphases and gene sequences. Univ. Texas Publ. **7213**: 163-177.

Dobzhansky, T. 1972. Species of Drosophila. Science **177**: 664-669.

Hardy, D. E. 1971. Evolution of the Hawaiian Drosophilidae (Insecta: Diptera). Symposium on Indian Ocean and Adjacent Seas, Abstract No. **228**, sec. xxii: 147-148.

Hardy, D. E. and Kaneshiro, K. Y. 1971. New picture-winged Drosophila from Hawaii, part II. Univ. of Texas Publ. **7103**: 151-170.

Hardy, D. E. and Kaneshiro, K. Y. 1971. New picture-winged Drosophila from Hawaii, Part III. Univ. Texas Publ. **7213**: 155-161.

Heed, W. B. 1971. Host plant specificity and speciation in Hawaiian Drosophila. Taxon **20** (1): 115-121.

Kambysellis, M. P. 1970. Compatibility in insect tissue transplantations. I. Ovarian transplantations and hybrid formation between Drosophila species endemic to Hawaii. *J. Exp. Zool.* **175** (2): 169-180.

Robertson, F. W. 1970. Evolutionary divergence in Hawaiian Drosophila. Sci. Prog. Oxford Univ. **58**: 525-538.

Rockwood, S. 1969. Enzyme variations in natural populations of Drosophila mimica. Univ. of Texas Publ. **6918**: 111-132. (not reported in 1970 review)

Rockwood, E. S., Kanapi, C. G., Wheeler, M. R. and Stone, W. S. 1971. Allozyme changes during the evolution of Hawaiian Drosophila. Univ. of Texas Publ. **7103**: 193-212.

Stains, H. J. 1971. Review, in Bioscience **21** (8): 393-394.

Stalker, H. D. 1970. The phylogenetic relationships of Drosophila species groups as determined by the analysis of photographic chromosome maps. Proc. XII Intern. Congress of Genetics **1**: 194.

Stalker, H. D. 1972. Intergroup phylogenies in Drosophila as determined by comparisons of salivary banding patterns. Genetics **70**: 457-474.

Wheeler, M. R. and Hamilton, N. 1972. Catalog of Drosophila species names, 1959-1971. Univ. of Texas Publ. **7213**: 257-268.

Wheeler, M. R. and Wheeler, L. 1972. Notes on some introduced Drosophila in Hawaii. Dros. Inf. Serv. **48**: 77.

Yoon, J. S., Resch, K. and Wheeler, M. R. 1972. Cytogenetic relationships in Hawaiian species of Drosophila. I. The Drosophila hystricosa subgroup of the 'modified mouthparts' species group. Univ. of Texas Publ. **7213**: 179-199.

Yoon, J. S., Resch, K. and Wheeler, M. R. 1972. Cytogenetic relationship in Hawaiian Drosophila II. The Drosophila mimica subgroup of the 'modified mouthparts' species group. Univ. of Texas Publ. **7213**: 201-212.

Yoon, J. S., Resch, K. and Wheeler, M. R. 1972. Intergeneric chromosomal homology in the family Drosophilidae. Genetics **71**: 477-480.

Patterns of speciation in Hawaiian *Drosophila* inferred from ancient chromosomal polymorphism

HAMPTON L. CARSON

Department of Genetics, University of Hawaii, Honolulu, Hawaii 96822

Polymorphisms due to chromosomal aberrations such as inversions or trans-locations are widely known in existing natural populations of many organisms, especially Diptera. Study of the allelic frequencies of the components of these polymorphisms, together with their geographical distributions, have indicated that some of them must have existed in the balanced state in populations for a considerable number of generations. How long this has been in the geological sense, however, is not clear, although in some cases, attempts at an estimate have been made (e.g. Mayr 1945; Stebbins 1945).

This paper will make use of the extensive data now available on inversions and their distribution in a number of species of Hawaiian *Drosophila* (Carson *et al.* 1970; Clayton, Carson and Sato 1972). Evidence will be adduced which indicates which of the two components of particular intraspecific polymorphisms is the ancestral arrangement of genes. This conclusion is built on both comparative cytological and zoogeographical data. These considerations permit not only an estimate of the geological age of certain chromosomal polymorphisms but also inferences about the patterns of speciation in past populations.

CAN INVERSIONS BE POLYPHYLETIC?

Any statement about the age of a chromosomal mutant may only be made with confidence if the mutant concerned can be identified as having resulted originally from an unique event, in other words, if it is monophyletic. Thus, if the aberration can arise more than once from an ancestral sequence, two observed mutant gene arrangements may indeed not be identical by descent.

Most point mutations and electrophoretically-detected variants appear to have appreciable mutation rates, on the order of 10^{-5}. Accordingly, the presence of observationally identical mutants in populations widely separated in space and time may indeed be due to the separate and independent (polyphyletic) origin of the mutants concerned.

With inversions and translocations, the situation appears to be different. Each such aberration results from two simultaneous breaks in the linear integrity of one or two chromosome arms. In those organisms having giant polytene chromosomes, the position of such breaks can be mapped with considerable accuracy by ordinary light microscopy. The degree of this accuracy is, however, limited by the observational techniques. Its reliability may be estimated in the following manner. In highly favorable *Drosophila* material, such as the picture-winged *Drosophila* of Hawaii, a polytene chromosome arm, such as X or 3 (Figure 1), displays about 150 easily resolvable chromatic discs. This number might be larger if very fine bands, which may be seen if the chromosome is stretched, were also included in the count. A conservative estimate, however, is deemed advisable for the following calculation.

If both the origin and survival of inversions in a chromosome arm is random, an inversion identical to one already found should arise in $(1/150)^2$ or $1/22,500$ cases. Only about 200 inversions have been described among the picture-winged *Drosophila* of Hawaii. That random origin and survival is not always indicated, however, has been pointed out by a number of authors (e.g. Stalker 1966). The various phenomena which appear to affect this situation can be illustrated using data on the large number of species of Hawaiian *Drosophila* in which inversions have been mapped.

Figures 1A and 1B show, respectively, the standard photographic salivary gland chromosome maps of chromosomes X and 3 of the species *Drosophila grimshawi*. The distal ends of the chromosome are to the left in each case. On the X chromosome are marked the positions of both breaks of each of 21 inversions (42 breaks) as observed in 68 species of the *D. grimshawi* subgroup and the closely related *D. punalua* subgroup. The breakpoints of each inversion are indicated by two similar lower-case letters. (In cases where a superscript is used, e.g. 'a²', an inversion wholly different from 'a' is indicated; the alphabet has been used several times in the labelling procedure.) On the third chromosome (Figure 1B), 16 inversions (32 breaks) are plotted.

All inversions included here are naturally-occurring and most are inversion differences which have become fixed between species. In a number of instances on both chromosomes there is correspondence of breakpoints. The most striking place where this is apparent is at the centromere end of each chromosome (right ends, Figures 1A and 1B). Such breaks are apparently located in the proximal heterochromatin of these acrocentric arms and the fact that many appear to correspond in position is misleading since their precise position cannot be mapped.

Each of these inversions with one break in the centromeric heterochromatin has also one break in the euchromatic section of the chromosome. Thus, diagnosis of such chromosomal mutants as unique is based on the observation of a single break only. In a number of instances on both chromosomes X and 3, however, correspondence of break positions occurs out in the apparent euchromatic section of the chromosome arm. Thus, in six places in chromosome X and three in chromosome 3, one of the two breaks of an inversion corresponds in position to one of the two breaks of another inversion. This is far more correspondence than

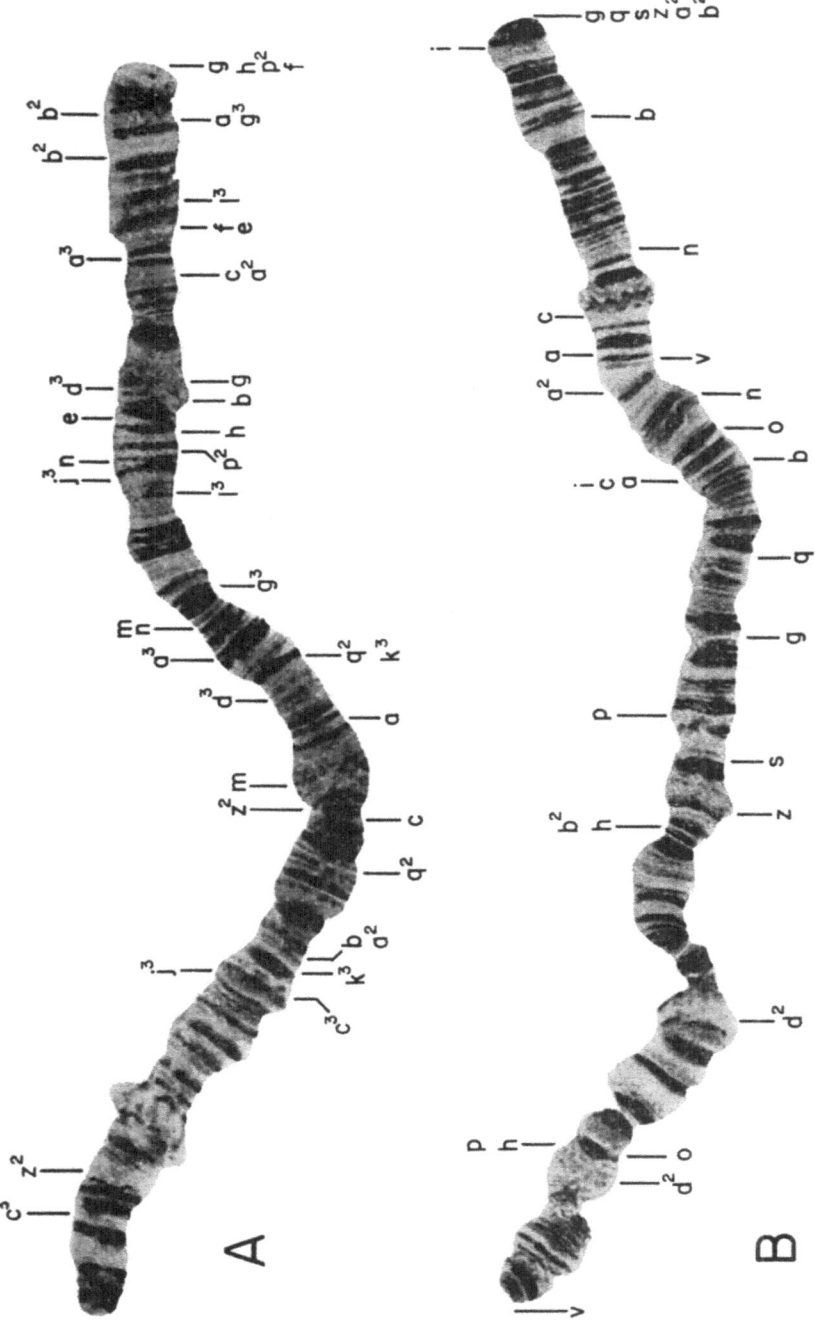

Figure 1. Standard polytene chromosome maps of the acrocentric chromosomes X and 3 of *Drosophila grimshawi*; the distal ends are to the left. The break points of 37 naturally-occurring inversions (21 in X and 16 in 3) are plotted as observed in 68 species.

would be expected on the hypothesis that the breaks in this part of the chromosome are random.

These correspondences may be due to the presence of linearly extensive but cytologically cryptic heterochromatic blocks located within the interstitial areas. In the instance of the point on chromosome 3 where 3a, 3c and 3i show corresponding breaks, there is cytological evidence for the existence of such an interstitial block; this region, for example, displays an apparent redundancy difference in hybrids between several of the species. The other correspondences appear to require another explanation.

In two of the instances on the X (m, n; f, e) and one in chromosome 3 (p, h) the correspondence involves the common break of two tandem inversions which are found in the cis position in natural populations. This suggests the not unreasonable hypothesis that these aberrations arose simultaneously and that only three breaks were involved in the formation of two inversions.

Additionally, before non-randomness may be accepted for the remainder of the correspondences, the possibility of selective survival of roughly similar aberrations must be considered. Thus Carson (1969) has presented evidence based on comparative cytology of chromosome 4 of the *D. grimshawi* subgroup which demonstrates the very great similarity of polymorphisms between closely related species. Thus, for example, the central section of chromosome 4 is involved in existing inversion polymorphisms in five different species. Close inspections show that the breakpoints are almost all different but the fact remains that much the same chromosome region is involved in each case. This leads to the conclusion that this region of the chromosome has selective value when heterozygous and that natural selection has favored and perpetuated randomly-arising inversions embracing this region. This results in parallel but not identical polymorphisms in the different species.

Finally, judgment that two breaks are cytologically identical should ideally be confirmed by examination of hybrids. In most cases, this has not been possible between many of these species, and non-correspondence may indeed exist for some breaks which are indicated as corresponding. It should further be remarked that even exact correspondence at the cytological level need not imply submicroscopic correspondence as well.

The question of whether there exist breakable sites in chromosomes or not has never been satisfactorily solved. The kind of data discussed above, dealing with naturally-occurring breaks, has sometimes been invoked as evidence that such places exist in chromosomes. On the other hand, the whole problem of differential survival of aberrations, discussed above, casts doubt on the existence of breakable sites.

Most evidence from breaks induced experimentally or by mutator genes or factors (e.g. Helfer 1941; Ives 1950; Levitan 1964) the break distributions approach randomness except for the heterochromatic areas. The limitations of cytological observations in these regions have already been discussed.

The possibility of enhanced breakability in or near heterochromatic regions, however, cannot be dismissed. Experimentally-induced inversions with one break in the proximal heterochromatin are frequently associated with variegation position effects and in some cases they are characterized by instability in the heterozygous state. Thus, Novitski (1946) has subjected the very long induced inversion rst[3] of the X chromosome of *D. melanogaster* to close scrutiny. In a number of cases, apparent reinversions from the heterozygous state were detected, attesting to the instability of the new sequence. Novitski claims that genetic tests indicate these reinversions to be precise. Nevertheless, because of the fact that the original proximal break is in the heterochromatin, both breaks of the reinversion could also be located in heterochromatic blocks and thus neither could be precisely mapped by either genetic or cytological methods.

Intricate relationship-diagrams of closely related species based on inversion-sharing have been published for two large arrays of species of *Drosophila* (Wasserman 1963 for the *repleta* group and Clayton, Carson and Sato 1972 for the Hawaiian flies). It is obvious from the above discussion that caution must be exercised in the use of inversions with one break in the heterochromatin or which manifest the phenomenon of parallel polymorphism.

The basic fact appears to remain, however, that when both breaks of an inversion can be mapped in the euchromatic sections of the giant chromosomes, the probability of separate and independent origin of an identical two-break aberration becomes vanishingly small. In the discussion which follows, no use has been made of inversions having one break in the heterochromatin.

ANCESTRAL AND DERIVED SEQUENCES

After an inversion has occurred, the population of the species contains two alternate arrangements of genes. Now if such polymorphism is the result of induction by some artificial means of mutagenesis, there is no difficulty in recognizing which of the two arrangements is the 'old', or ancestral, one and which is the 'new' or recently-induced. This is, of course, providing that a standard gene order has been observed and established prior to treatment.

When a polymorphism is found in a natural population, however, the oldness of one of the alternatives cannot be recognized by any intrinsic property of one of the two arrangements of genes. Diagnosis of the ancestral character of one of the two gene arrangements must be established by use of some other kind of information.

The purpose of the present paper is to discuss several cases of polymorphism in Hawaiian *Drosophila* in which the ancestral nature of one of the two gene arrangements can be determined by consideration of comparative sequential cytology between species. Coupled with data on geographical distribution of the species and the geological age of the islands concerned, inferences about past events of speciation can be drawn. A terse account of this case has been published (Carson 1973).

CHROMOSOMAL POLYMORPHISM IN DROSOPHILA NEOPICTA

This species of picture-winged Hawaiian *Drosophila* is found on East Molokai and West and East Maui (Figure 2). Although East and West Maui volcanos were at one time in the geologic past separated by a water barrier, they are now part of the same island, being joined by a low saddle. All three volcanos are young geologically; potassium-argon dating of the lava flows of East Molokai indicate that it is the oldest of the three, approximately 1.5 million years (MacDonald and Abbot 1970). East Maui, the giant dome of Haleakala, shows no lava flows older than one million years.

D. *neopicta* is a member of the *planitibia* subgroup of species (Carson and Stalker 1968; Clayton, Carson and Sato 1972). The grouping of species has been made on the basis of the particular fixed inversions which the species share. Sixteen species are now recognized in the subgroup. Of these, fourteen, including D. *neopicta*, are unusual among *Drosophila* in having an extra crossvein in cell R-5 of the wing. D. *neopicta* is a rather small fly but some of its relatives are among the largest, if not the largest, in the entire genus.

The *planitibia* subgroup, along with eighty other picture-winged species, has been completely described chromosomally in terms of the polytene sequences displayed by the Standard, *Drosophila grimshawi*. In the course of this basic phylogenetic work, small samples of D. *neopicta* were obtained. The species does not breed well in the laboratory and unlike most of the Hawaiian picture-winged *Drosophila*, the females appear to oviposit better when they are kept in small groups rather than as isolated specimens.

The chromosomal polymorphisms found in these samples and their distributions over the three volcanos are given in Table 1. The cytological extent of the

Table 1. Chromosomal polymorphisms in *Drosophila neopicta*. Frequencies (in per cent) of each gene arrangement in three localities.

Gene arrangment:	Locality:		
	Molokai	West Maui	East Maui
X	50	0	0
Xv^2	0	100	86
Xt	50	0	0
Xtw^2	0	0	14
2	25	0	39
2m	75	0	0
2mn	0	100	61
4	50	100	100
$4f^3$	37	0	0
$4a^3g^3$	13	0	0
No. of autosomes observed (N_A)	8	10	18
No. of X chromosomes observed (N_X)	6	9	14

various inversions present have been described previously; inversion Xt shares one break in common with inversion Xs (Carson and Stalker 1968). The data on the distribution of these inversions in *neopicta* are few and should be viewed only as qualitative information as to the presence or absence of the specified gene arrangement.

Several points are of interest. The X/Xt polymorphism is present on both Molokai and East Maui. On Maui, the homologue carrying the X arrangement always is found with the Xv^2 inversion, that is, it is completely linked with it in a cis position. Xt on Maui is similarly always accompanied by another inversion, Xw^2. On Molokai, these latter inversions have not been found and a simple X/Xt polymorphism prevails.

Table 1 also gives data on the autosomal polymorphisms of this species. 2/2m also exists on both Molokai and East Maui, although in all Maui populations, the 2m arrangement is accompanied by a second inversion, 2n, which overlaps 2m. A 4th chromosome polymorphism, $4/4f^3$, exists on Molokai but has not been found on Maui. On Molokai, a single homologue, otherwise having the 4 arrangement, was found to have two other inversions designated as $4a^3$ and $4g^3$. In the discussion which follows, three polymorphisms (X/Xt, 2/2m and $4/4f^3$) are particularly stressed. The distribution of these gene arrangements in *D. neopicta* populations is summarized in Figure 2.

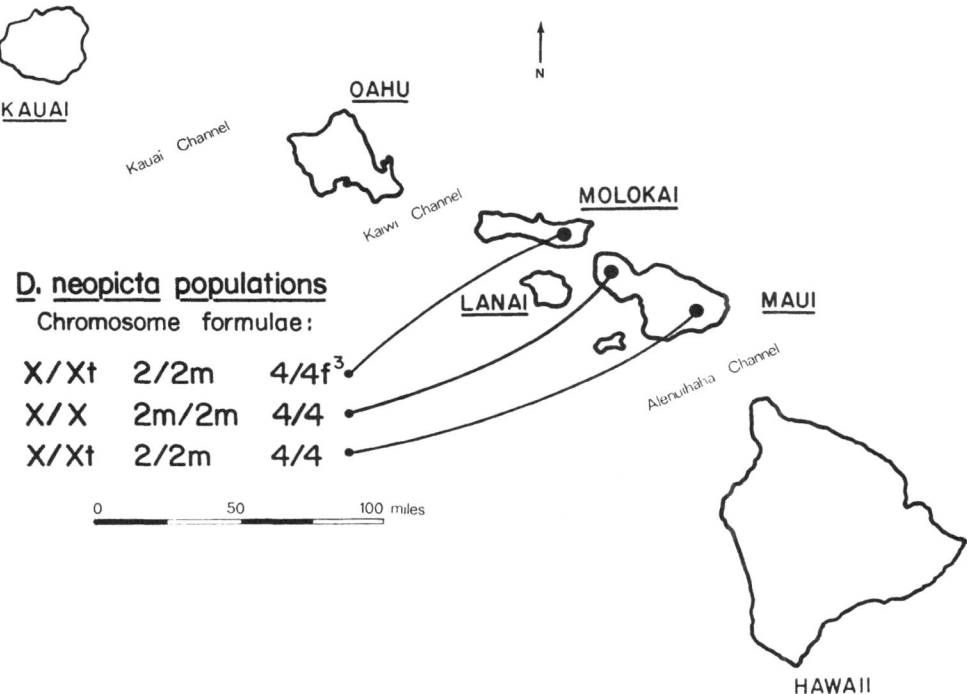

Figure 2. Distribution of *Drosophila neopicta* and its chromosomal polymorphisms in chromosome X, 2 and 4.

The comparative cytological information on the picture-winged *Drosophila* summarized by Clayton, Carson and Sato (1972), embraces 96 species and 198 inversions. A striking feature of these data is the fact that it is possible to determine the distribution of any given two alternate arrangements of bands throughout this large series of species. A study of the distribution of X and Xt yields the results given in Table 2. Thus, the arrangement called X (lacking Xt) is found to the exclusion of its alternative in 84 species. Xt alone is found in eleven species; only *neopicta* is polymorphic.

Among those species lacking Xt are the known members of subgroup V, *D. primaeva* and *D. attigua* (Carson and Stalker 1969). The banding patterns of these species have been shown to be closer to those of certain continental

Table 2. Distribution of the two gene arrangements comprising the X/Xt polymorphism of *D. neopicta* in 95 related Hawaiian species.

Gene arrangments present	Distribution in subgroup or species	No. of species	No. of chromosomes examined
X only	I. grimshawi	56	2075
	III. adiastola	14	330
	IV. punalua	8	303
	V. primaeva	2	68
	II. planitibia (in part):	4	
	picticornis		113
	setosifrons		15
	substenoptera		5
	obscuripes		9
		84	2918
X, Xt	*neopicta*	1	29
Xt only	II. planitibia (in part):	11	
	cyrtoloma		16
	hanaulae		3
	hemipeza		15
	heteroneura		15
	ingens		2
	melanocephala		11
	neoperkinsi		10
	nigribasis		12
	oahuensis		15
	planitibia		36
	silvestris		44
		11	179

species (the *D. robusta* group in particular) than any other of these 96 Hawaiian *Drosophila* (Stalker 1972). Accordingly, the species which lack Xt are judged to be more primitive than the eleven species in which Xt is fixed. These facts strongly indicate that of the two components of the polymorphism, as observed in present-day *neopicta* populations, X is the older, or ancestral arrangement, and Xt is the newer, or derived one.

The distribution of one or the other of the two components of a second and a fourth chromosome polymorphism in *D. neopicta* can also be traced in other species in a similar manner. The polymorphisms involved are 2/2m and 4/4f³. Parallelism with the X chromosome polymorphism is striking. Thus the same 84 species which have X (Table 2) also lack 2m and 4f³; this is based on the

Table 3. Summary of the distribution of ancestral (X, 2, 4) and derived (Xt, 2m, 4f³) gene arrangements in the species of the *planitibia* subgroup.

Gene arrangements present			Species	Geographical distribution	Number of chromosomes 2 and 4 examined
X	2	4			
X	2	4	*obscuripes*	Maui	12
			picticornis	Kauai	146
			setosifrons	Hawaii	20
			substenoptera	Oahu	8
X, Xt	2, 2m	4, 4f³	*neopicta*	Molokai, E. and W. Maui	36
Xt	2, 2m	4, 4f³	*neoperkinsi*	Molokai	12
Xt	2m	4f³	*ingens*	W. Maui	4
			melanocephala	E. Maui	14
Xt	2m	4	*nigribasis*	Oahu	16
Xt	2	4	*cyrtoloma*	E. Maui	18
			hanaulae	W. Maui	4
			hemipeza	Oahu	18
			heteroneura	Hawaii	20
			oahuensis	Oahu	16
			planitibia	E. and W. Maui	46
			silvestris	Hawaii	58

examination of 3859 representatives of autosome 2 and 4. Table 3 gives the detailed distribution of the ancestral (X, 2, 4) and derived (Xt, 2m, 4f³) arrangements in the sixteen species of the *planitibia* subgroup.

Four of the species lack any of the derived arrangements. Two of these, *picticornis* (Kauai) and *setosifrons* (Hawaii) are, in terms of their polytene sequences, much more primitive than the other fourteen (Clayton, Carson and Sato 1972). These two, furthermore, are the only species of this subgroup which lack the 'extra' crossvein in cell R-5 of the wing. The other two species which lack derived arrangements are *obscuripes* (East Maui) and *substenoptera* (Oahu). Both of these are chromosomally very close, in fact, homosequential with standard *D. neopicta*; which of these two is closest to the putative ancestor of the extra-vein flies cannot be estimated from these data.

Figure 3 shows the geographical distribution of the species of the members of the *planitibia* subgroup (other than *D. neopicta*) which show one or more

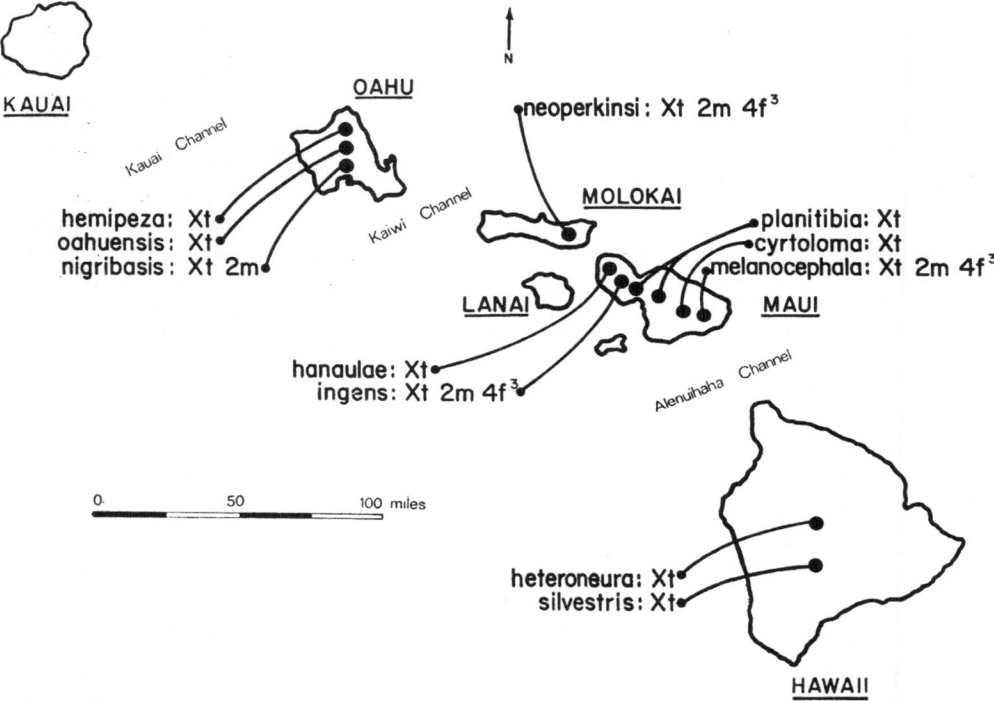

Figure 3. Distribution of the members of the *planitibia* subgroup of Hawaiian Drosophila (other than *neopicta*) which display one or more derived chromosomes (i.e. Xt, 2m, 4f³).

of the three pertinent derived chromosome arrangements. It will be recalled that *neopicta* is found on three volcanos, Molokai and East and West Maui. All the rest of these species, with the apparent exception of *D. planitibia* (*sensu stricto*) on East and West Maui, are endemic to single volcanos.

DISCUSSION

The conditions in modern *D. neopicta* populations are of special interest. Populations of this species from the island of Molokai carry old/new polymorphisms in three chromosome pairs. *D. neoperkinsi,* a Molokai endemic sympatric with with *neopicta,* has the same two autosomal polymorphisms but has apparently fixed the derived X arrangement. Only these two species carry the ancestral along with derived arrangements as a polymorphism; the ten other species showing derived arrangements carry them in the fixed state only. The conclusion may be drawn that the polymorphisms, as found in *neopicta* populations of Molokai, are ancient, having persisted since before the time that the species with the fixed derived arrangements were formed. Based on the age of the islands involved this would appear to be between one and one-half and two million years.

The question may be raised whether old/new polymorphisms as found in *D. neopicta* and *D. neoperkinsi* might not be due to recent hybridization between monomorphic species rather than direct descent from an ancient polymorphism. This is considered to be extremely unlikely because of the fact that *neopicta* is an unusual species of the *planitibia* subgroup; there is no species on Maui or Molokai which is close to it in size or morphology. Progeny of single wild-caught females of all of these species maintain their distinct morphological characters in laboratory cultures with great precision with no hint of variation in the direction of sympatric species. *D. neopicta,* moreover, is alone among the Maui and Molokai flies not only in its small size but in its poorly developed lek behavior (H. T. Spieth, personal communication). This renders copulations with other species unlikely although this possibility should be tested in the laboratory.

All of the facts of the case suggest that all twelve of the existing species which carry derived gene arrangements were ultimately descended from a singly-, doubly-, or triply-polymorphic ancestral population. Doubly- and triply-polymorphic species populations exist today in *neoperkinsi* and *neopicta* on Molokai. The facts further suggest a pattern of speciation in which a large, widespread ancient population, carrying considerable variability, produces allo-patric species on its margins. In the Hawaiian situation this appears to result in descendent species with small populations and restricted distributions. Wasser-man (1963), Throckmorton (1966) and Stalker (1972) have presented inferences on past speciation events from consideration of conditions in descendent species. In each case, long-standing heterozygosity is suggested for the ancestral species. The facts presented here are wholly consistent with these ideas and with the marginal speciation concept presented by the author some years ago (Carson 1959). Thus, it may be suggested that certain large, longstanding, widespread populations may be especially prone to produce small, isolated, inbred populations on their margins. These may become species through peripheral isolation and the founder effect (Carson 1971). Certain large, long standing populations may be judged to be especially competent to produce species in the manner described. Its new products, which may come to coexist with it, may not necessarily inherit the competence of their parent species. Thus, there is little evidence, in this case, that the species descended from *neopicta*-like ancestral populations are themselves equally competent to continue the process. Indeed, many of them may well have reduced variability, specialized adaptations and small populations. Such features would appear to restrict their future evolutionary potential; they may be referred to as incompetent for speciation. More data on the comparative population genetics of such close clusters of species is much to be desired.

SUMMARY

The Hawaiian species *Drosophila neopicta* carries a number of intra-specific inversion polymorphisms. Utilizing comparative cytological, geographical and geological evidence, it is possible to identify, in three instances, which of the two components of a chromosomal polymorphism is ancestral and which derived. One of the derived arrangements is fixed in populations of eleven species. The polymorphism retained in modern *neopicta* must accordingly be older than these species, on the order of one-and-a-half to two million years. The reliability of the inversion data in such inferential procedures is discussed and the suggestion is made that some populations may be more competent than others to produce clusters of new allopatric species.

ACKNOWLEDGEMENT

This work has been supported by grants GB 27586 and GB 29288 from the National Science Foundation.

REFERENCES

Carson, H. L. 1959. Genetic conditions which promote or retard the formation of species. Cold Spring Harb. Symp. Quant. Biol. **24**: 87-105

Carson, H. L. 1969. Parallel polymorphisms in different species of Hawaiian *Drosophila*. Amer. Nat. **103**: 323-329.

Carson, H. L. 1971. Speciation and the founder principle. Stadler Symposia **3**: 51-70.

Carson, H. L. 1973. Ancient chromosomal polymorphism in Hawaiian *Drosophila*. Nature **241**: 200-202.

Carson, H. L., Hardy, D. E., Spieth, H. T. and Stone, W. S. 1970. The Evolutionary biology of the Hawaiian Drosophilidae. In: Essays in Evolution and Genetics in Honor of Theodosius Dobzhansky. M. K. Hecht and W. C. Steere, Eds. Appleton-Century-Crofts, N.Y. pp. 437-543.

Carson, H. L. and Stalker, H. D. 1968. Polytene chromosome relationships in Hawaiian species of *Drosophila*. II. The *D. planitibia* subgroup. Univ. Texas Publ. **6818**: 355-365.

Carson, H. L. and Stalker, H. D. 1969. Polytene chromosome relationships in Hawaiian species of *Drosophila*. IV. The *D. primaeva* subgroup. Univ. Texas Publ. **6918**: 85-94.

Clayton, F. E., Carson, H. L. and Sato, J. E. 1972. Polytene chromosome relationships in Hawaiian species of *Drosophila*. VI. Supplementary data on metaphases and gene sequences. *Univ. Texas Publ.* **7213**: 163-177.

Helfer, R. G. 1941. A comparison of x-ray induced and naturally occurring chromosomal variations in *Drosophila pseudoobscura*. Genetics **26**: 1-22.

Ives, P. T. 1950. The importance of mutation rate genes in evolution. Evolution **4**: 236-252.

Levitan, M. 1964. The first thousand aberrations produced by a maternal factor: a progress report. Genetics **50**: 265-266.

Mayr, E. 1945. Age of the distribution pattern of the gene arrangements in *Drosophila pseudoobscura*. Some evidence in favor of a recent date Lloydia **8**: 70-83.

Novitski, E. 1946. The regular reinversion of the roughest[3] inversion. Genetics **46**: 711-717.

Stalker, H. D. 1966. The phylogentic relationships of the species in the *Drosophila melanica* group. Genetics **53**: 327-342.

Stalker, H. D. 1972. Intergroup phylogenies in *Drosophila* as determined by comparison of salivary banding patterns, Genetics **70**: 457-474.

Stebbins, G. L. Jr. 1945. Age of the distribution pattern of the gene arrangements in *Drosophila pseudoobscura*. Evidence for abnormally slow rates of evolution with particular reference to the higher plants and the genus *Drosophila*. Lloydia **8**: 84-102.

Throckmorton, L. H. 1966. The relationships of the endemic Hawaiian Drosophilidae. Univ. Texas Publ. **6615**: 335-396.

Wasserman, M. 1963. Cytology and phylogeny of *Drosophila*. Amer. Nat. **97**: 333-352.

Mating behavior and evolution of the Hawaiian *Drosophila* [1]

HERMAN T. SPIETH

University of California, Davis, California 95616

The endemic Hawaiian Drosophilidae fauna exhibits a number of unique features which differentiate it from the rest of the world's species. A partial list of such features includes (1) the large size of many of the species, (2) striking pigmentation patterns of numerous species, and especially the prevalence of patterned wings, (3) numerous diverse, often bizarre sexual dimorphic characters, (4) a high level of agonistic behavior exhibited both intra- and inter-specifically, (5) complex courtship patterns, and (6) the existence of intermediates between the genera *Scaptomyza* [2] and *Drosophila* (Hardy 1970, Spieth 1970, Throckmorton 1966).

All of these features are directly or indirectly related to the courtship behavior and evolution of the rich fauna packed into the Hawaiian forest, especially the montane rain forests which cover only a minor portion of the 6318 square miles of area of the six Hawaiian islands on which the flies live. Although considerable areas of the native forests have been decimated by Western culture, it is clear that even before the Polynesians arrived the forests and the associated *Drosophila* were able to inhabit only limited portions of the islands.

A few *Drosophila* species from various other parts of the world have minor sexual dimorphisms which are usually restricted to the male fore tarsi, e.g., the sex combs of *melanogaster, simulans, pseudoobscura, persimilis* and their relatives. Occasionally the color pattern of the abdominal tergites is slightly different between the two sexes. In the Hawaiian flies, however, the great majority of the males are dimorphic and almost any part of the body may be involved, e.g., antennae, mouthparts, head shape, setal pattern of head, forelegs involving not only the tarsi but also the tibia and femur, the wings, and various segments of the abdomen. In addition to the morphological dimorphisms, there are accom-

1 The research upon which this paper was based was supported in part by grants from the National Science Foundation and the National Institute of Health to D. E. Hardy of the University of Hawaii and Wilson S. Stone of the University of Texas.
2 In the present discussion, the endemic *Scaptomyza* species are excluded from detailed consideration.

panying pigmentary dimorphisms involving the mouthparts, face, antennae, wings, legs and one or more abdominal segments (see Hardy 1965, 1970). The number of dimorphisms per individual species varies, usually two to four; in some more than four and rarely only one. *What is invariable about the male dimorphisms is the fact that each and every one of them is involved in the male's courtship behavior.*

If one dissects the male courtship pattern of a non-Hawaiian species such as *melanogaster,* it is found to consist of five elements: two types of leg vibration; pulses of wing vibration by one wing; licking of the female's ovipositor; circling about her body. After displaying, the male then attempts mounting and intromission. Few, if any, non-Hawaiian species have more complex behavior than does *melanogaster* but some, e.g., *auraria,* have simpler courtships, involving only wing action, which occurs as the male mounts and seeks to achieve intromission.

In comparison, most of the Hawaiian species of the genus *Drosophila* exhibit complex courtship patterns and few are as 'simple' as that of *melanogaster.* As an example, *truncipenna* males exhibit ten clearly discrete courtship elements, consisting of two foreleg, two mouthpart, two wing and three abdominal movements plus circling. Further, its wing displays involve *both* wings simultaneously, instead of a simple wing vibration as in *melanogaster.* The *truncipenna* male has dimorphic color marking on its forelegs, face, mouthparts, wings and abdomen. These are displayed to the female as the male engages in a contorted posture immediately in front of her face. Such color dimorphisms are rare in non-Hawaiian species.

Hawaiian *Drosophila* species also display three other mating behavior features that are lacking in non-Hawaiian species. These are (1) the lek territories of the males, (2) immobility of copulating pairs and (3) catalepsy of the males of some species at the termination of copulation.

Put briefly, males and females of non-Hawaiian species diurnally visit food sites during morning and evening. At the food sites, the females devote most of their time and attention to feeding and ovipositing. The males spend only a short time feeding and then persistently approach and attempt to court and mate with various females. *In toto* there is a considerable amount of overt activity on such a food mass.

In comparison, the Hawaiian drosophiloids do not engage in courtship on the food sites; rather both sexes are extremely quiescent and tolerant of each other's presence, restricting themselves solely to feeding activities. Both sexes arrive at, and slip away from, the food site by short quick flights close to the substrate or slow deliberate walking. After each short flight the individual remains immobile for a period before resuming its approach or departure.

In the vegetation surrounding and relatively near the food sites, the sexually mature males select and individually defend a lek territory. Typically each lek territory is small: an upper or lower surface of a leaf; a short section of a fern frond stem or a tree limb; a smooth spot, several inches in diameter, on a large limb or trunk of a tree. Males of most species select locations from which they may intercept females as they leave the food site and return to their reposing areas in the surrounding vegetation. Each male having established his lek territory then engages in advertising actions which serve to attract sexually receptive females. Advertising behavior is species specific but can be divided into three types: (1) waving both wings and walking alertly about on the lek territory (visual advertising); (2) sitting immobile with the tip of the abdomen elevated and pulsating an anal droplet; (3) walking about and dragging the tip of the abdomen over the substrate and depositing a thin film of liquid on the substrate. The last two types of activity result in the release of a volatile pheromone into the atmosphere. All three types of behavior have been observed in both the laboratory and in the field.

If a female is attracted to the male's lek by his advertising and if copulation results following the subsequently occurring courtship, then as soon as the union is formed the pair becomes immobile and remains so until the individuals physically separate. Under laboratory conditions with a number of individuals in an observation cell, a courting pair is alert to the near presence of another individual and usually the courtship will be broken off by the close approach of a third fly. Once copulation occurs and the pair becomes immobile, then another individual can literally walk on top of the pair without any response being given. In comparison, copulating pairs of non-Hawaiian species have females that walk about carrying the males and fending off by leg and wing actions any nearby individuals.

Near the end of the copulatory period the male of some Hawaiian species, e.g., *crucigera* and others, releases his legs from grasping the female and assumes a cataleptic posture. As soon as genitalic withdrawal occurs, the male falls to the substrate, sometimes dropping a considerable distance, and then often lies immobile for one or two minutes before quickly jumping to an upright position. Even in species in which catalepsy does not occur, the physical separation is accomplished with scant movement on the part of either individual.

Individuals of non-Hawaiian species show variable levels of agonistic behavior. Both males and females fend against each other with their legs, especially the middle legs, and a few species such as *virilis* and its relatives engage in a ritualized sidewise arching and sharp thrusting against each other, an action termed curling (Spieth 1952). In comparison, most Hawaiian species display a much higher level of agonistic behavior. Practically all species engage in curling but in addition a common ritualized behavior occurs which I call slashing. The aggressive individual first orients upon a nearby or approaching fly, then scissors both wings outward and horizontally from its body in increasing amplitude, and just as the wing motions reach 90°-100° from the median line he or she raises, thrusts and fully straightens the forelegs forward and upwards and then sharply slashes the stiff legs downward. The first slashing action often occurs when the opponent is at such a distance that the legs do not strike its body and just the occurrence of the action will often cause the intruder to retreat or move away. If the intruder is not deterred, then both flies slash repeatedly at each other until one flees.

A more spectacular agonistic behavior called upright fighting is displayed by some of the *planitibia* subgroup species. These are the largest *Drosophila* known with body and wing lengths up to 7.0-8.0 mm, and the behavior seems restricted to males who are competing for a lek territory. Two males approach each other head-on and gradually raise their heads upward by extending the forelegs but at the same time keeping the tips of their abdomens close to the substrate. As they slowly move close to each other, raising their heads higher with repeated bobbing movements, they reach a point where the longitudinal axis of their bodies forms an angle of about 70°-80° with the substrate, and their long forelegs are almost fully extended. When their heads are about 0.5 cm apart, each raises and then carefully places its middle legs in front of the forelegs, turns its head slightly away from the opponent, then raises its forelegs and wraps them around the opponent simultaneously, thrusting heads and thoracic venters against each other. The mouthparts are stiffly extended and used as a lever to try to depress the other individual downward. The two interlocked flies may struggle in this upright position for several minutes before one is defeated. In both upright fighting and the slashing actions, the lowering of the anterior end of the body towards the substrate is a signal of submission. If two flies of unequal size approach each other, the smaller one is never able to bob its head higher than the larger one, and as a result it turns and flees before actual physical interlocking occurs.

The males of all species are agonistic to any outsider entering their lek territory. If a female enters she is treated agonistically until the male determines her sex; then he courts but if she is non-receptive he drives her away. Males often intrude onto the leks of nearby neighbors and conflict invariably results but the outcome is unpredictable except that one fly always departs. Species that have abandoned lek behavior and the Hawaiian *Scaptomyza* species which apparently never exhibited lek behavior are much less agonistic, paralleling the non-Hawaiian species in their level of agonistic behavior.

The general behavior of the Hawaiian species in their normal habitats can best be described as secretive, wary and cryptic. The flies never hover over food sites as do individuals from other areas. Except for rare instances (such as the litter-dwelling *mimica*) when disturbed, they do not fly upward but rather by a rapid darting flight dive downward toward the substrate and into areas of lower light intensity. They respond sharply to passing shadows and to the quick movements of any object about them but remain immobile if approached slowly. Finally, they choose substrates to rest upon with which their color patterns blend effectively, at least to the human observer.

Thus upon the basic life style of the genus, evolutionary processes in the Hawaiian fauna have resulted in the development of an extraordinary complex of interrelated behavioral and morphological features which include the large size of many species, behavioral and morphological crypsis, the spatial separation of adult feeding and courtship, diverse and numerous sexual dimorphisms, complex courtships and agonistic behaviors. It is patent that these evolutionary innovations did not occur accidentally or by mere chance on this small group of isolated mid-Pacific islands. Strong selection pressure or pressures must have been in operation.

Before considering what these selection pressures might be, it is necessary to consider briefly the ovipositional choices of the females and resultant larval substrates. Heed (1968, 1971) has investigated these matters extensively, and Montgomery (1972) has effectively expanded these investigations. Heed has reared more than 180 drosophiloid and scaptomyzoid species from fermenting parts of 40 plant genera. His data show that fallen fermenting leaves are the commonest larval substrate. Rotting bark is next in importance, followed in decreasing order by fermenting flowers, fruits, fungi, fluxes and frass. More interesting is the fact that 50 species were bred from fermenting leaves of *Cheirodendron* (Araliaceae), 25 others from the leaves of the lobeliad *Clermontia* (Campanulaceae), and 15 from *Ilex* (Aquifoliaceae) leaves. When the numerous bark, flower and fruit breeders of *Cheirodendron* and *Clermontia* are added to the 75 leaf breeders on the two plants, it is clear that these two genera provide the larval substrates for the majority of the species. Study of the various species groups shows that the more derived groups utilize leaves, especially those of *Cheirodendron*, and that the more primitive species groups are concentrated on the bark, leaves and fruits of *Clermontia* and the related lobeliad genus *Cyanea*. Also, the fermenting parts of the lobeliads are prime food sources for a great many adults, including a number that do not use it as a larval substrate. Large numbers of individuals, therefore, accumulate around these plants.

The honey-creeper birds constitute the dominant vertebrates of the native fauna and until the destructive impact of Western culture, they existed in vast numbers (Perkins 1903, 1913). Many species of honey-creepers are, like the drosophiloids, ecologically intertwined with the lobeliads, especially the *Clermontia* and *Cyanea* species, since the birds avidly feed upon the nectar of the flowers, and this action secondarily makes them the prime pollinators of the lobeliads.

Additionally, the birds feed upon insects and Warner (1967) placed field-captured Iiwi (*Vestaria coccinea*) in a large cage and offered them three species of native Drosophilidae of varying sizes. The birds quickly pursued and captured the insects but significantly the wing buzzing of the largest species, *D. grimshawi*, caused the birds to release them while they avidly ate the medium and small sized flies. Burger (personal comm.) exposed similar insects to a honey-creeper that had been reared on an artificial diet in the laboratory and it also pursued and ate the insects. When collecting the flies by smearing bait on tree trunks at Kokee, Kauai, I repeatedly had groups of the amakihi (*Chloradrepanis virens virens*) find and continually raid the bait sites, leaving their beak marks on the bark as they struck at the feeding drosophiloids. In addition, the small endemic flycatcher or Elepaio (*Chasiempsis sandwichensis*) is constantly present in the lower layers of the vegetation whenever drosophiloids are numerous. Feeding on insects it catches them on the wing and also with its wren-like behavior picks them from the bark and leaves of the plants.

In addition to the birds, numerous species of the predatory muscoid genus

Lispocephala are abundant in the same habitats where the drosophiloids live. The *Lispocephala* oviposit in the same material as do the *Drosophila* and their carnivorous larvae feed avidly on the drosophiloid larvae. The adults have been observed both in the field and in the laboratory to be capable of capturing and eating small to medium sized adult drosophiloids but they can not capture the larger species. Whenever numerous *Drosophila* adults are present in the forests, then also *Lispocephala* individuals are numerous.

I therefore suggest that the insectivorous behavior of the honey-creepers, the flycatcher, and the *Lispocephala* species has served as a powerful selection pressure upon the Hawaiian *Drosophila*. This selection pressure resulted in four discrete but interrelated evolutionary modifications of the biology of the flies: (1) the selection of adults that are alert, wary and highly cryptic, both behaviorally and structurally; (2) the drastic increase in size of some species, especially those that tend to live in relatively exposed sites in the forests; (3) the abandonment of courtship on the feeding and ovipositional site thus eliminating the incessant courtship activity that could be quickly detected by the birds and the *Lispocephala* adults; (4) the emergence of lek behavior and a concomitant drastic increase in sexual selection.

It is to be noted that the flies have no control over the location of the food sites and that they are often found in situations that clearly expose them to predators. Crypsis then becomes the main protective strategy. In comparison the flies select their leks and typically these are situated so as to avoid predation. Further, although a number of males usually have leks in relatively close proximity, the flies are not congregated as closely together as they are on the small food substrate areas.

With courtships restricted to the lek, the male can no longer move randomly from female to female in his courting as do the non-Hawaiian species but must attract a female singly to his small lek. Even when she is attracted to his lek, his subsequent courtship must then induce her to mate and a male displaying a complex and precisely unique courtship pattern involving a high level of stimulation (i.e. super-stimuli) gains advantage.

The male exhibiting the highest level of agonistic behavior has an advantage since the lek must be kept free of other males and also of any female who might be attracted but then refuses to copulate. Additionally, as soon as a copulation is finished, the female must be driven away to make room for another possible sexual candidate.

Thus the evolution of the lek behavior provides the selection pressure that has resulted in the high level of agonistic behavior and the complex male courtship displayed by the males while the female responses to the male's courtship still retain the primitive characteristics of all *Drosophila* species with one clear exception, i.e., the immobility during copulation which reflects the direct effect of predatory activities upon the flies.

In sum, most major aspects of the flies' behavior are tailored to the avoidance of predation.

Not all Hawaiian species of *Drosophila* utilize lek behavior and members of the small subgenus *Engiscaptomyza,* which have the external morphology of a drosophiloid but the behavior, some internal anatomy and the egg structure of a scaptomyzoid, do not utilize lek courtship. Rather, as the flies move about in the forest, when a male comes into close proximity to another individual he simply lunges onto it, grasping its body with his large hypertrophied forelegs and then curls the tip of his abdomen downward and seeks to achieve intromission. If a receptive female has been mounted, copulation occurs. If instead another male or a non-receptive female has been mounted, then after a time he dismounts.

Only six species belong to the subgenus *Engiscaptomyza:* four in the *crassifemur* subgroup, two in the *nasalis* subgroup. The four *crassifemur* subgroup species all have essentially similar external morphology and can be effectively separated only on the basis of the male genitalia (Kaneshiro 1969). Also only a single species is present on each of the major islands with the Maui complex of Maui, Molokai and Lanai (Carson and Stalker 1968) sharing a single species. The related *nasalis* subgroup has two species, one on the Maui complex and one on the island of Hawaii. Thus allopatric inter-island speciation has occurred but on each island the gene flow has been sufficient so that no intra-island speciation has occurred.

In comparison, numerous examples can be identified in the lek species where a single mountain top has served as a sufficient area to allow the isolation and evolution of a species. For example, on the Maui complex three lek species (*hamifera,* East Maui volcano; *paenehamifera,* West Maui volcano; and *varipennis,* East Molokai volcano) all are closely related even to the extent that they have homosequential polytene chromosomes, yet are morphologically and behaviorally distinct.

A more complex example is found in the well studied *planitibia* subgroup. One section of the subgroup consists of three closely related species, *neopicta, obscuripes,* and *substenoptera.* The last two species are homosequential and are thus parallel chromosomally to *hamifera* and its relatives. *D. obscuripes,* the most primitive of the trio, is restricted to East Maui and displays lek behavior. In comparison, both *neopicta* and *substenoptera* have abandoned lek behavior and find and court their females in a fashion basically similar to that of *E. crassifemur* and its relatives. *D. neopicta* has also a similar geographical distribution to that of *E. crassifemur,* being found on both East and West Maui and Molokai. *D. substenoptera* is found only on Oahu and represents the descendants of a migrant from the Maui complex, presumably from the same ancestral stock that gave rise to the present day *neopicta.*

In sum, lek behavior with its concomitant heightened sexual selection apparently results in isolated populations which occupy relatively and absolutely small geographical areas being able quickly to achieve reproductive isolation. Non-lek species under similar situations do not develop effective sexual isolating mechanisms since, under the lower levels of sexual selection that pertain with the typical drosophiloid and scaptomyzoid courtship, behavioral changes in the courtship patterns occur much more slowly.

I would also note that behavioral evidence indicates that speciation in the

Hawaiian drosophiloids has occurred allopatrically and that not one example exists to date that would suggest the occurrence of sympatric speciation.

In conclusion, I would emphasize that a major factor in the evolutionary development of the Hawaiian drosophiloids has been the simple fact that the ancestral drosophiloid immigrant to the islands invaded an ecological niche which perforce exposed it to effective and persistent predation. Coupled with the physical nature and geographical arrangement through time of the various Hawaiian islands, this predation was a major driving force in the evolution of a rich and diverse fauna.

REFERENCES

Carson, H. L. and H. D. Stalker. 1968. Polytene chromosome relationships in Hawaiian species of *Drosophila*. I. The *D. grimshawi* subgroup. Univ. Texas Publ. **6818**: 335-354.

Hardy, D. E. 1965. Insects of Hawaii, Vol. 12. Diptera: Cyclorrhapha II, Series Schizophora, Section Acalypterae I. Family Drosophilidae. Honolulu: Univ. Hawaii Press, 814 pp.

Hardy, D. E. 1970. Drosophilid Fauna in Carson, H.L. et al. The Evolutionary Biology of the Hawaiian Drosophilidae. Essays in Evolution and Genetics in Honor of Theodosius Dobzhansky. New York: Appleton-Century-Crofts, 450-468 pp.

Heed, W. B. 1968. Ecology of the Hawaiian Drosophilidae. Univ. Texas Publ. **6818**: 387-420.

Heed, W. B. 1971. Host specificity and speciation in Hawaiian *Drosophila*. Taxon **20** (1): 115-121.

Kaneshiro, K. Y. 1969. The *Drosophila crassifemur* group of species in a new subgenus. Univ. Texas Publ. **6918**: 79-84.

Montgomery, S. L. 1972. Host selection by picture winged *Drosophila*. Master's Thesis, Univ. of Hawaii.

Perkins, R. C. L. 1903, 1913. Fauna Hawaiiensis: Vol. 1, Pt. 4, Vertebrata; Vol. 1, Pt. 6, Introductory essay on the fauna. Cambridge Univ. Press.

Spieth, H. T. 1952. Mating behavior within the genus *Drosophila* (Diptera). Bull. Amer. Mus. Nat. Hist. 99 (7): 395-474.

Spieth, H. T. 1970. Biology and Behavior in Carson, H.L. *et al*. The Evolutionary Biology of the Hawaiian Drosophilidae. Essays in Evolution and Genetics in Honor of Theodosius Dobzhansky. New York: Appleton-Century-Crofts, 469-489 pp.

Throckmorton, L. H. 1966. The relationships of the endemic Hawaiian Drosophilidae. Univ. Texas Publ. **6615**: 335-396.

Warner, R. E. 1967. Some observations on the birds of Kipahulu Valley. Scientific Report of the Kipahulu Valley Expedition, 2 August–31 August 1967—Unpublished.

Phylogenetic relationships of Hawaiian Drosophilidae based on morphology

KENNETH Y. KANESHIRO

Department of Entomology, University of Hawaii, Honolulu, Hawaii 96822

INTRODUCTION

Concerning problems in the systematics of Drosophilidae, Sturtevant (1939) states that '... it is desirable that there be available a satisfactory arrangement of species into some scheme of classification that can be taken as indicating their degree of genetic relationships.' He discussed in detail the methods in analyzing and selecting characters which are important in showing phylogenetic relationships between species. It is clear that Sturtevant's description of taxonomic methodology in the classification of the genus *Drosophila* more than 30 years ago has remained a most important contribution in the biosystematics of *Drosophila*.

There are several lines of evidence which show that the present taxonomic status of the Hawaiian Drosophilidae may give a misleading interpretation of the phylogenetic relationships between species. This group of species has undoubtedly undergone explosive evolutionary radiation in a geologically short period of time. Currently there are approximately 500 described species (the endemic fauna may ultimately total upwards of 800 species) divided into eight endemic genera. Throckmorton (1966), based on a comparative study of the internal anatomy, states that there are only two major lineages in the evolution of Hawaiian Drosophilidae: the 'drosophiloids' and the 'scaptomyzoids.' Corroborating observations on the mating behavior patterns by Spieth (1966) show that there are two basic behavior patterns in the Hawaiian drosophilids: a very elaborate, species-specific courtship in the drosophiloids and a simple 'assault' courtship in the scaptomyzoids. Based on observations of the metaphase karyotypes, Clayton (1966, 1968) summarizes that the endemic species of Hawaiian Drosophilidae fall into two chromosomal groups which correspond with the genus *Drosophila* and the genus *Scaptomyza*. Recently, Yoon *et al.* (1972), based on a comparison of the polytene chromosomal patterns, show that there is a high degree of chromosomal homology between the genus *Drosophila* and the genus *Antopocerus* and conclude that these two genera have a common ancestor.

In this paper, there will be a brief discussion in the use of male genitalic structures as an important tool in studying the phylogenetic relationships of the species in the Hawaiian fauna. There will also be a discussion of how an analysis of various isolating mechanisms is important in the taxonomic treatment of the endemic drosophilids.

PHYLOGENETIC RELATIONSHIPS BASED ON GENITALIC STRUCTURES

Snodgrass (1957) states: 'The great diversity in structural detail of the genitalia gives these organs a value for the identification of insect species almost equal to that of fingerprints for identification of human individuals.' In most of the continental drosophilids, a comparative study of male genitalic structures plays an important role in showing species differentiation. There are many examples of pairs or groups of *Drosophila* species (e.g., *melanogaster* and *simulans, pseudoobscura* and *persimilis*, etc.) which are very similar and/or indistinguishable based on external morphological characters other than the genitalia; but for the most part, these 'sibling species' have been shown to be readily distinguishable based on structures of the male genitalia. As A. H. Sturtevant (1919) recognized over 50 years ago, the external male genitalia of Drosophilidae can be used as an important taxonomic tool in distinguishing between closely related species.

A comparative study of the phallic organs of the picture-winged species group of Hawaiian *Drosophila*, however, shows that there is very little structural diversity between closely related species (Kaneshiro, 1969). In most cases, one cannot distinguish between species within a species subgroup based on a comparison of phallic structures. It became clear then, that a study of the genitalic structures of the picture-winged species group could be used as an important tool in showing phylogenetic relationships between species, but that the phallic structures cannot be used to differentiate between closely related species. Based on genitalic characters, the picture-winged group of species is divided into nine species subgroups (Kaneshiro, 1969) which, for the most part, parallel very closely the phylogenetic relationships shown by cytological (Carson *et al.*, 1967, 1968a, 1968b, 1968c, 1969; Carson, 1971; and Clayton *et al.*, 1972), biochemical (Johnson, W. E., unpublished data), ethological (Spieth, 1966) and ecological (Heed, 1968 and Montgomery, 1972) data. In a few cases, a comparison of genitalic characters provides supplemental information as to the true relationships of species which are chromosomally 'homosequential' (Carson *et al.*, 1967), i.e., having the same banding patterns on all five long arms of the polytene chromosomes. This situation is illustrated in the case where *vesciseta* (Figure 1C) is shown to be chromosomally homosequential with *pilimana* and *glabriapex* (Figures 1A and 1B respectively) but is found to be more closely related to *virgulata* and *hexachaetae* (Figures 1D and 1E respectively) on the basis of genitalic characteristics. Carson and Stalker (1968a: 344) show that *vesciseta* differs from *virgulata* by one fixed chromosomal inversion and from *hexachaetae* by two fixed inversions.

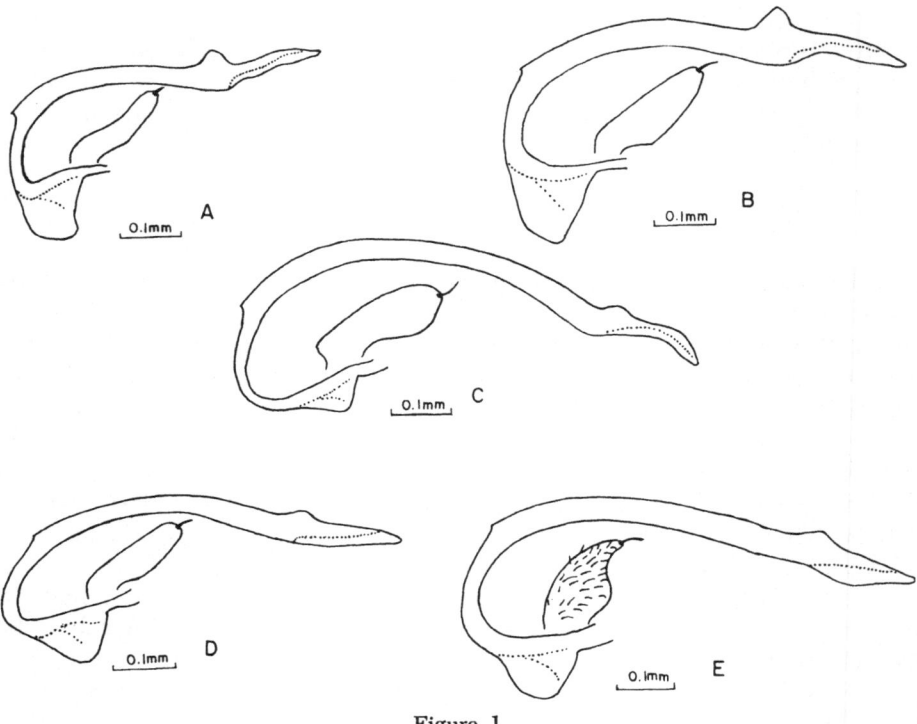

Figure 1

Based on the data on the picture-winged species group, it is possible to make comparisons of the genitalic structures of the other major groups in the endemic fauna and show phylogenetic relationships of the generic groupings. The original taxonomic treatment of the endemic species divide the fauna into nine endemic genera. A preliminary investigation of the male genitalia of representatives of the nine genera show that the species grouped in the various genera are probably only species complexes or at the most subgeneric groupings of only two genera: the genus *Drosophila* and the genus *Scaptomyza*. It became clear that a biosystematic study of Hawaiian Drosophilidae based strictly on external morphological characters can give a misleading impression of evolutionary divergence.

In the drosophiloids, which include the genera *Drosophila, Antopocerus, Nudidrosophila, Ateledrosophila* and what was formerly the genus *Idiomyia,* the various characters used to differentiate between the genera can be shown to be not important as generic characters. The species previously described in the genus *Idiomyia* Grimshaw are characterized by the presence of an extra crossvein in cell R5 in the wings of both sexes (Figure 2D). This is a very distinctive character and would appear to be a bona fide generic character. However, Carson *et al.* (1967) presented chromosomal evidence which indicate that the genus *Idiomyia* is congeneric with the genus *Drosophila*. They showed that two of the species, *clavisetae* and *neogrimshawi*, which have the extra crossvein in

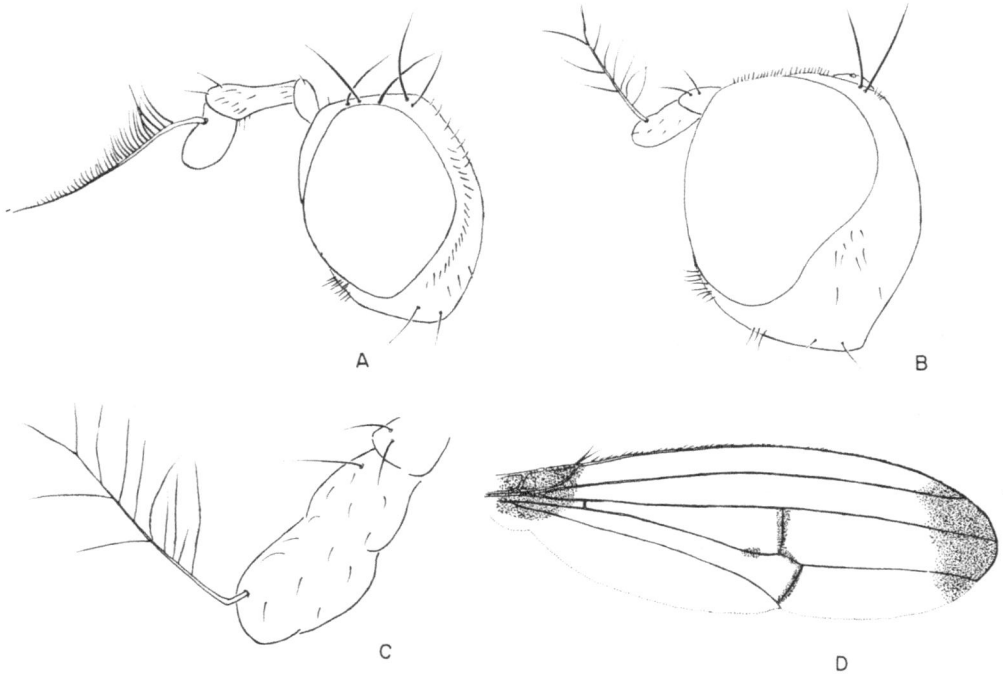

Figure 2

cell R5, differed from several species (i.e., *adiastola* subgroup species) which are typical *Drosophila* species by only five fixed inversions. The other 'idiomyia' species are 25+ inversions away from *clavisetae* and *neogrimshawi*. The phallic structures of *clavisetae* and *neogrimshawi* (Figures 3C and 3D respectively) are shown to be similar to species in the *adiastola* subgroup (Figures 3A and 3B) as reported by Kaneshiro (1969) while the phallic structures of the remaining idiomyia species (e.g. Figures 3E and 3F) are distinctly different and more closely resemble those of *setosifrons* and *picticornis* (Figures 3G and 3H respectively). It is apparent that the extra crossvein in cell R5 arose independently in the two lineages and therefore is not a valid generic character.

The key characters of the remaining three drosophiloid genera are mainly secondary sexual structures found only in the males and are structures which are used in the elaborate courtship behavior patterns (Spieth, 1966). The endemic genus *Antopocerus* Hardy consists of nine described species and is characterized by the large porrect antennae of the males (Figure 2A). The genus *Nudidrosophila* Hardy consists of five described species and is characterized by the males lacking the orbital and ocellar bristles and having microscopic pubescence or setae on the front (Figure 2B). The genus *Ateledrosophila* Hardy consists of two described species and is characterized by the males having the arista preapical in position (Figure 2C) and also by lacking the orbital and ocellar bristles. In most cases, the females of the species in these three genera cannot be distinguished

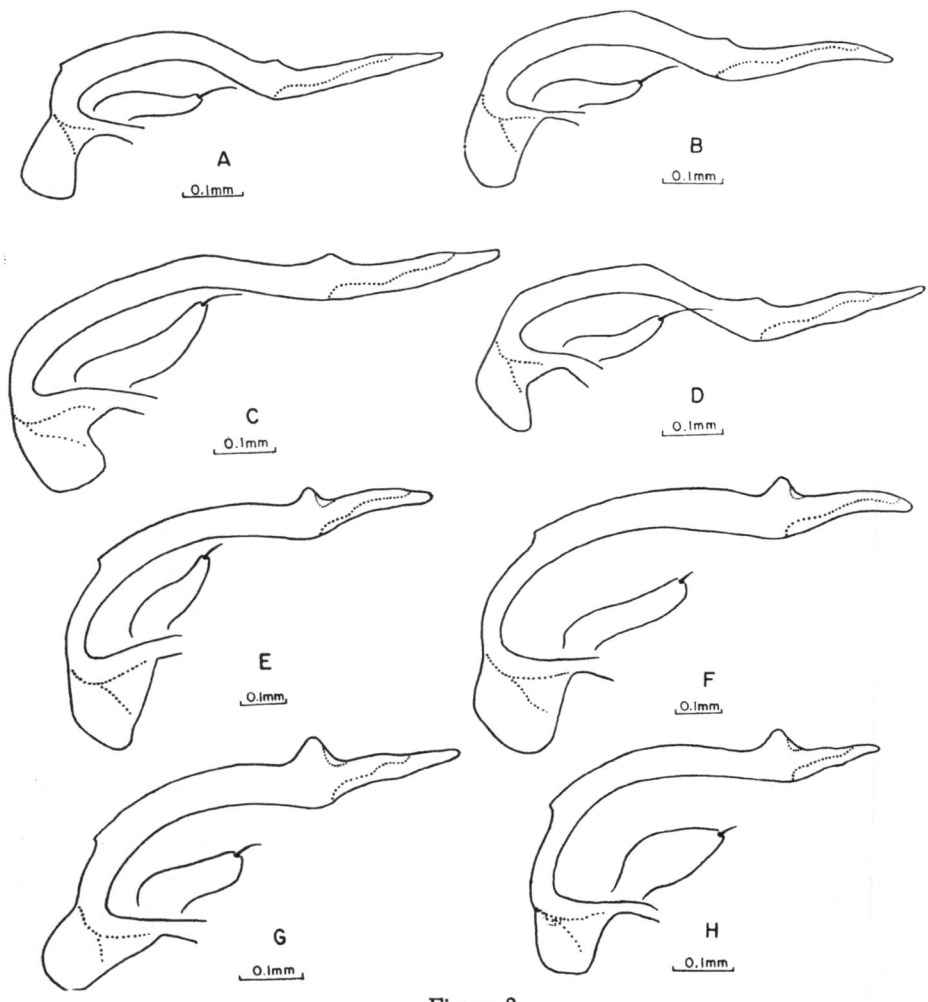

Figure 3

from females of typical *Drosophila* species. Also, there are *Drosophila* species which have phallic structures which are very similar to those of the species in these three genera. The modified tarsi group of species (Figures 4D, 4E, and 4F), for example, have very similar phallic organs as those of the *Antopocerus* species (Figures 4A, 4B and 4C). The aedeagus of *Nudidrosophila* species has a characteristic hook-like structure near the apex (figs. 5A and 5B) and for a time, it appeared that this character could possibly be used as a good generic character. However, it was found that there are two species in the genus *Drosophila* which have this same hook-like preapical structure on the aedeagus (figs. 5C and 5D). It is evident, then, that the conventional morphological characters which most Diptera taxonomists might have used as key generic characters are not alway reliable when studying the phylogenetic relationships of Hawaiian drosophiloids.

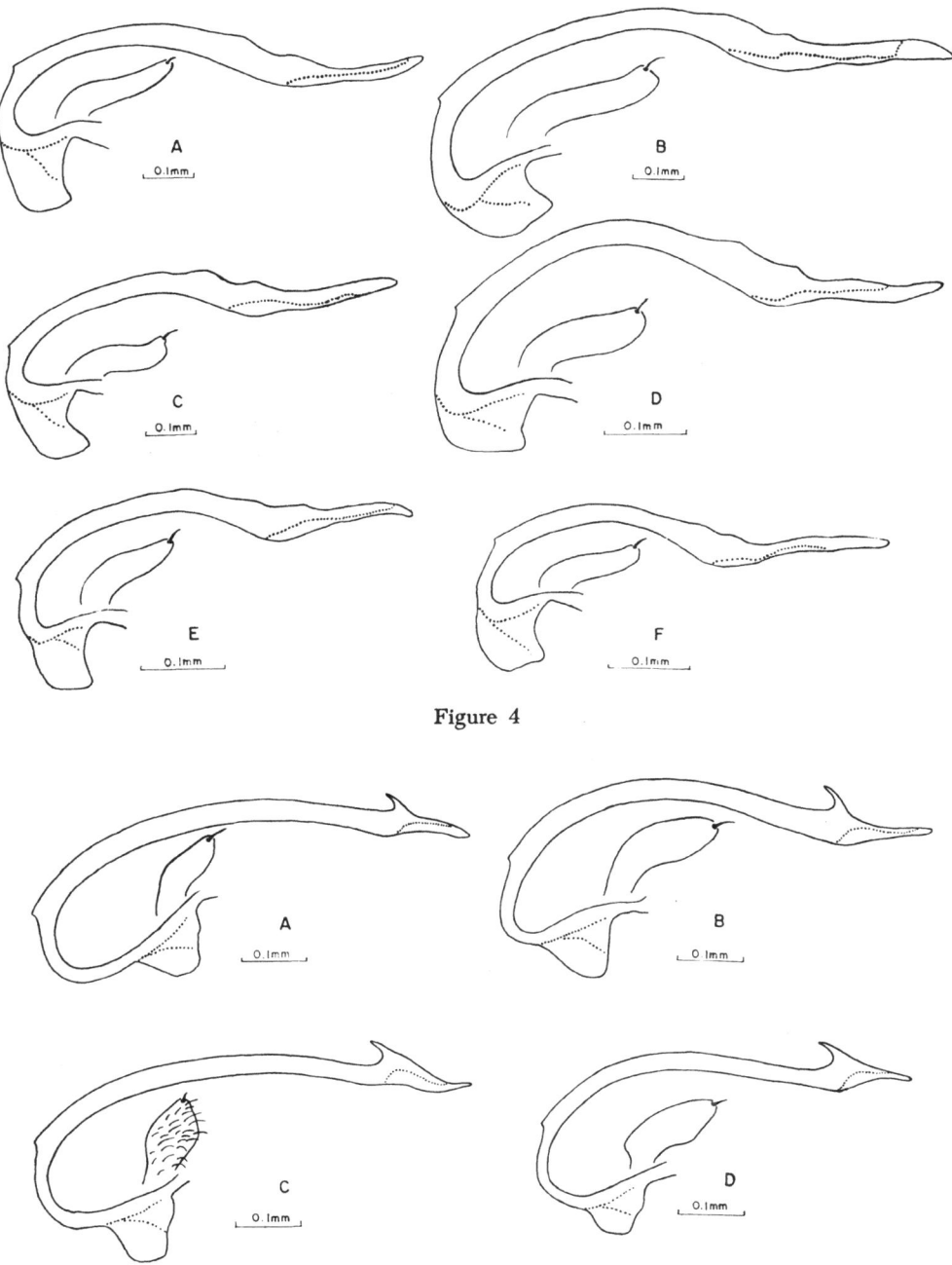

Figure 4

Figure 5

The data on the scaptomyzoid group of species which include the genera *Scaptomyza, Celidosoma, Grimshawomyia,* and the spider egg parasite, *Titano-*

chaeta, is incomplete so that this group will not be discussed at length at this time. However, it is observed that the scaptomyzoids are exactly the opposite of the Hawaiian drosophiloids in that lek behavior is absent and mating behavior is relatively simple. Therefore, there has not been any selection for the development of secondary sexual structures as in the drosophiloids and there is very little diversity in the external morphology between closely related species. The scaptomyzoids have instead evolved a highly complex genitalic apparatus which can be used to differentiate between closely related species.

THE 'SIBLING SPECIES' OF HAWAII

Another important taxonomic problem we have encountered in our evolutionary study of the drosophilid fauna of Hawaii is in the concept of 'sibling species.' The classical definition of 'sibling species' is a group of species which are morphologically similar or identical natural populations that are reproductively isolated (Mayr, 1971). *D. pseudoobscura* and *persimilis, melanogaster* and *simulans,* and *mulleri* and *aldrichi* are classical examples of sibling species pairs in *Drosophila.* In all of these cases, the external morphology, excluding the male genitalic structures, is nearly identical. However, there are good differentiating characters in the genitalic apparatus so that one could easily separate *pseudoobscura* from *persimilis, melanogaster* from *simulans* and *mulleri* from *aldrichi.* Now, in the evolution of Hawaiian *Drosophila,* elaborate courtship and mating behavior has played a very important role as a premating isolating mechanism between species (Spieth, 1966). Therefore, there is a high degree of diversity in the external morphology which is manifested mainly by the secondary sexual structures found only in the males of the species. The females of closely related species are usually indistinguishable. For the most part, even one who may be totally unfamiliar with the *Drosophila* fauna of Hawaii may be able to readily distinguish between males of closely related species. However, the phallic structures of closely related species are very similar and in most cases, indistinguishable. This is a complete reversal of the situation found in the continental sibling species where you have little or no morphological diversity but usually with good genitalic differences; whereas in Hawaiian *Drosophila* you have tremendous morphological diversity but with little or no genitalic differences.

There are preliminary evidence from some unpublished hybridization experiments that sibling species in the classical sense do exist in the Hawaiian fauna. All of these, however, are pairs or groups of species which are morphologically extremely close even in genitalic characteristics and are found on separate islands. In these cases, geographic isolation has apparently played the major role in reproductive isolation so that there has not been selection for differentiation of the secondary sexual structures found in sympatric species pairs. To the present, only one case of sympatric sibling species pair as described by the classical definition of sibling species has been found in Hawaii. *D. primaeva* and *attigua* occur sympatrically in isolated areas on the island of Kauai and can be differentiated by only minor differences in the phallic structures.

It is evident then, that if one were to determine and to take into consideration the actual isolating mechanisms which are operating between species, the concept of sympatric sibling species could be broadened to include some of the closely related species of Hawaiian *Drosophila* even though the species are morphologically readily distinguishable.

SUMMARY AND CONCLUDING REMARKS

It is obvious that the endemic fauna of Hawaiian Drosophilidae represents a classical example of evolutionary radiation on oceanic islands. This remarkable group of species has speciated profusely over a relatively short span of geological history. The data accumulated by various investigators based on cytology, behavior, electrophoresis, ecology and morphology indicate that the endemic drosophiloids make up a group of closely related species despite the tremendous morphological diversity that is so characteristic of this unique fauna. Furthermore, the data presented by comparisons of the male genitalic apparatus show that even species originally described in separate genera are only species complexes or at the most subgeneric groupings of the genus *Drosophila*. The key generic characters which are conventionally used by most Diptera taxonomists are apparently of no phylogenetic importance in the biosystematic study of Hawaiian drosophiloids. It is clear that a careful study of the biological factors responsible for reproductive isolation between species is essential. In the Hawaiian drosophiloids, the evolution of an elaborate, species-specific courtship and mating behavior has played an important role in the reproductive isolation between species. This type of premating isolating mechanism is strongly reflected by the tremendous diversity in the external morphology of these species. Therefore, these key 'adaptive' characters which have been used to differentiate between species and also to group species into separate genera are not important in showing phylogenetic relationships between species or group of species.

On the basis of the evidence presented by the various investigators discussed above and the comparisons of the male genitalic structures, it is clear that the drosophiloids of Hawaii represent a very close knit group of species which probably evolved from a common ancestor. Therefore, it is probable that the genera *Antopocerus* Hardy, *Nudidrosophila* Hardy, and *Ateledrosophila* Hardy are congeneric and should be synonymized with the genus *Drosophila*. A more detailed presentation of the various evidence which document this conclusion is being considered in a dissertation by the author.

It is quite clear that the Drosophilidae of the Hawaiian Archipelago presents taxonomists with a group of animals which is undoubtedly speciating at an accelerated rate and taxa at all stages of incipient speciation can be found. A careful study of these species and an analysis of their speciation mechanisms are crucial in the understanding of the evolutionary process. In the study of the evolution and genetics of Hawaiian Drosophilidae, various basic concepts in the field of systematics and taxonomy are being tested and with the efforts of a team of investigators in various fields of biology, new concepts will undoubtedly be formulated.

REFERENCES

Carson, H. L., Clayton, F. E. and Stalker, H. D. 1967. Karyotypic stability and speciation in Hawaiian *Drosophila*. Proc. Nat. Acad. Sci., 57(5): 1280-1285.

Carson, H. L. and H. D. Stalker, 1968a. Polytene chromosome relationships in Hawaiian species of *Drosophila*. I. The *D. grimshawi* subgroup. Univ. Texas Publ. 6818: 335-354.

Carson, H. L. and H. D. Stalker. 1968b. Polytene chromosome relationships in Hawaiian species of *Drosophila*. II. The *D. planitibia* subgroup. Univ. Texas Publ. 6818: 355-366.

Carson, H. L. and H. D. Stalker. 1968c. Polytene chromosome relationships in Hawaiian species of *Drosophila*. III. The *D. adiastola* and *D. punalua* subgroups. Univ. Texas Publ. 6818: 367-380.

Carson, H. L. and H. D. Stalker. 1969. Polytene chromosome relationships in Hawaiian species of *Drosophila*. IV. The *D. primaeva* subgroup. Univ. Texas Publ. 6918: 85-94.

Carson, H. L. 1971. Polytene chromosome relationships in Hawaiian species of *Drosophila*. V. Additions to the chromosomal phylogeny of the picture-winged species. Univ. Texas Publ. 7103: 183-192.

Clayton, F. E. 1966. Preliminary report on the karyotypes of Hawaiian Drosophilidae. Univ. Texas Publ. 6615: 397-404.

Clayton, F. E. 1968. Metaphase configurations in species of the Hawaiian Drosophilidae. Univ. Texas Publ. 6818: 263-278.

Clayton, F. E., Carson, H. L. and Sato, J. E. 1972. Polytene chromosome relationships in Hawaiian species of *Drosophila*. VI. Supplementary data on metaphases and gene sequences. Univ. Texas Publ. 7213: 163-178.

Heed, W. B. 1968. Ecology of the Hawaiian Drosophilidae. Univ. Texas Publ. 6818: 387-419.

Heed, W. B. 1971. Host plant specificity and speciation in Hawaiian *Drosophila*. Taxon 20: 115-121.

Kaneshiro, K. Y. 1969. A study of the relationships of Hawaiian *Drosophila* species based on external male genitalia. Univ. Texas Publ. 6818: 55-70.

Mayr, E. 1970. Populations, Species, and Evolution. Belknap Press of Harvard Univ. Press, Cambridge, Mass.

Montgomery, S. L. 1972. Master's Thesis. Department of Entomology, Univ. of Hawaii.

Snodgrass, R. E. 1957. A revised interpretation of the external reproductive organs of male insects. Smithsonian Misc. Collection 135(b): 1-11.

Spieth, H. T. 1966. Courtship behavior of Hawaiian Drosophilidae. Univ. Texas Publ. 6615: 245-313.

Sturtevant, A. H. 1919. A new species closely resembling *Drosophila melanogaster*. Psyche 26: 135-155.

Sturtevant, A. H. 1939. On the subdivision of the genus *Drosophila* Proc. Nat. Acad. Sci. 25: 137-141.

Throckmorton, L. H. 1966. The relationships of the endemic Hawaiian Drosophilidae. Univ. Texas Publ. 6615: 335-396.

Yoon, J. S., Resch, K. and Wheeler, M. R. 1972. Intergeneric chromosomal homology in the family Drosophilidae. Genetics 71: 477-480.

Degrees of reproductive isolation between closely related species of Hawaiian *Drosophila* °

ELYSSE M. CRADDOCK

Department of Genetics, University of Hawaii, Honolulu, Hawaii, U.S.A. and Department of Population Biology, Research School of Biological Sciences, The Australian National University, Canberra, A.C.T., Australia

I INTRODUCTION

It is a basic tenet of evolutionary thought that the gene pool of every fully differentiated species is discretely isolated from that of any other species (Dobzhansky 1970; Mayr 1970). Diverging populations must therefore acquire some means of reproductive or genetic isolation from one another in the course of species formation. If such isolation is to be fully effective, the mechanisms responsible must be an inherent property of the species: this means that they must be genetically based, rigidly canalized, and preferably operative at several levels. Further, unless the genes which guarantee reproductive isolation are fixed into the genotype of every individual of the species, gene flow between the diverging entities remains possible, and any genetic differentiation which may have arisen is potentially reversible. The acquisition of substantial reproductive isolation is thus a critical point in the speciation process, and one which determines to some extent whether speciation progresses to completion. Obviously, this is not an instantaneous event, any more than that involved in the acquisition of morphological or other differences associated with species divergence. It should therefore be possible to study the evolution of reproductive barriers, and this might best be done in an actively evolving group which includes forms at various levels and stages of divergence. Furthermore, the results of such a study should provide a direct means of assessing the mode or modes of speciation involved in the radiation of the group of organisms concerned.

The biological features of the endemic drosophilid fauna of the Hawaiian Islands make these flies uniquely suited to evolutionary studies. Speciation in the group has been extraordinarily prolific, and Hardy estimates that it may

° Supported by N.S.F. Grant GB-29288 for Evolution of Hawaiian Drosophilidae Project

include some 650 to 700 species (Carson *et al.* 1970). Even more remarkable than this fact is the necessarily recent and rapid nature of the speciation events, indicated by the comparative geological recency of the Hawaiian Islands (McDougall 1964; Stearns 1966). Many of the present species must be very young in evolutionary terms. None of the species of the Big Island of Hawaii, the most south-easterly island in the chain, can be more than 700,000 years old, the estimated age of the oldest volcano on the Island (McDougall 1969; Dalrymple 1971); and most of them would probably be considerably younger than this. Due to the movement of volcanic activity from the north-west to the south-east, and the sequential development of the islands in this direction, there must have been a time differentiation in the colonization of the various islands by *Drosophila*. The extremely high frequency of single-island endemism in the group (Carson 1971) implies that many of the speciation events occurred in association with or subsequent to colonization events. Hence, it is possible to infer considerable variation in the time of speciation of various members of the group.

The picture-winged group of Hawaiian drosophilids offer another advantage for the comparative study of levels of reproductive isolation. This closely knit group comprises 110 species which are readily recognized by their diverse and highly distinctive morphological characteristics (Hardy and Kaneshiro 1972). Significantly, the general pattern of phylogenetic relationships within the group has been established independently of knowledge of the reproductive barriers. These relationships have been derived from an analysis of chromosomal inversion sequences in the polytene salivary gland chromosomes (Carson and Stalker 1968*a*, *b, c*, 1969; Carson *et al.* 1970; Carson 1971). Consideration of data from morphological, geographical and geological studies gives direction to the chromosomal phylogeny, and permits an interpretation of the ancestral-derivative relationships within this phylogeny. A complete picture of the temporal sequence of species differentiation requires some assumptions regarding the number of founding events and the directions taken by inter-island founders (*cf.* for example Carson 1970), but these assumptions are reasonable, and the scheme of relationships obtained seems to be biologically meaningful. Within the picture-winged group then, there is a good range of species of varying ages and varying degrees of evolutionary divergence. Some populations may even represent primary stages of divergence, as undoubtedly speciation is still in progress in the group. They thus present an ideal situation in which to study species relationships, degrees of isolation, and the development of the species gap itself.

The mechanisms which ensure reproductive isolation and maintain two populations as distinct entities or species can be extremely varied. In any given situation, the more important question is not 'what are the isolating mechanisms?', but rather, 'how did they originate?'. However, description must come before interpretation. This paper therefore deals with:—

(1) the nature of the reproductive barriers found between species of the picture-winged group of Hawaiian drosophilids

(2) the evolution of these barriers, and

(3) speciation patterns within the group.

II REPRODUCTIVE ISOLATING MECHANISMS IN DROSOPHILA

The concept of reproductive isolation is used here in its widest sense to denote the overall genetic isolation between two biologically distinct systems. In essence, genes from one gene pool are prevented from dispersing freely into a foreign gene pool, and vice versa, by the specific isolating devices which are the basic property of each and every species (Dobzhansky 1951, 1970; Mayr 1942, 1963). An important and useful distinction has been made between the various kinds of isolating mechanism. With premating isolation, the barrier is imposed prior to actual reproductive contact, most often by ecological or sexual separation of the two forms. With postmating isolation, the barrier operates at a later stage, as a consequence of events in the actual products of reproductive contact. The times of action of the various kinds of reproductive barriers can thus be ordered into a precise sequence. Most often, the barriers to free exchange of genes between a pair of species are multiple ones, and usually, they involve both premating and postmating devices. The earliest-acting one necessarily provides the effective block to genetic exchange, and the later-acting mechanisms must be largely unused, serving merely as reinforcing mechanisms. At the extreme, in the case of allopatric species, the real functional deterrent to gene exchange is a physical geographic barrier, and not a genetic mechanism. In nature then, many species barriers are in a sense potential, and do not normally contribute to functional reproductive isolation. Laboratory investigation of these reserve mechanisms, by removal, breakdown or bypassing of the earliest-acting and operational barrier, often emphasizes the extent of the genetic gulf between species, and may indicate important phases in the process of species divergence.

Within the genus *Drosophila*, there are examples known of almost every possible kind of isolating device (*cf.* Patterson 1942; Patterson and Stone 1952; Dobzhansky 1970 for reviews). Some of these are more frequently involved in the effective isolation function than others. In particular, sexual or ethological isolation is widespread between sympatric species, so that natural hybridization is rare. Where premating barriers are lacking, isolation may still be effectively maintained by postmating barriers. In *Drosophila,* these are mainly dependent on hybrid sterility, which may be genic, chromosomal or cytoplasmic (Ehrman 1962). Hybrid inviability at various stages of development is also important in some cases. In general, premating barriers are assumed to be more efficient, since gamete wastage is avoided: they might also be assumed to be characteristic of older species which have been subject to evolutionary processes for long periods of time.

Analysis of the total nature of the reproductive isolation between two species must involve assessment of the geographic, ecological and ethological factors involved, in addition to estimation of the contributions made by hybrid inviability, hybrid sterility, or other forms of hybrid breakdown. Once the earlier-acting mechanisms have been detected, other means will usually be required to expose the later-acting mechanisms. In the first instance, experimental hybridization studies can provide substantial information on the components of isolation likely to be important in a given situation. But this approach is subject to some limitations. For example, laboratory culture may demonstrate the existence of prematation barriers without identifying their nature. On the other hand, laboratory culture may cause breakdown, so that postmating barriers such as sterility or inviability of hybrids are revealed, whilst the premating barriers are not manifest. Field studies and other approaches must then be used to supplement the laboratory hybridization data. Laboratory and field studies are complementary— both are necessary for a complete understanding and interpretation of the reproductive isolation between particular pairs of species. As far as possible in this paper, the genetic barriers between Hawaiian *Drosophila* species have been assessed on this basis. The available data on interspecific hybridizations are not extensive, and are confined to members of the picture-winged group. Most of the hybridization data for this review are drawn from Yang and Wheeler (1969), Kaneshiro (unpublished observations) and Craddock (in preparation): some additional information is based on other more limited studies and sporadic records of hybridizations collected by various workers in the course of the Hawaiian *Drosophila* Project. In a few instances where premating isolation is absolute, and the regular hybridization method inapplicable, the technique of ovarian transplantation between species has been successfully used to bypass the premating barrier and demonstrate the existence of postmating barriers (Kambysellis 1970).

The nature of reproductive barriers between Hawaiian species

The laboratory production of hybrids between Hawaiian *Drosophila* species has been extraordinarily successful, considering the extent of morphological and behavioral differentiation which is present. Yang and Wheeler's study (l.c.) used 28 species of the picture-winged group, and 20 of these were involved as parents of F_1 hybrid progeny in the successful crosses (16 per cent of those attempted). A large proportion of the crosses failed to give progeny. Strong sexual isolation is prevalent between many Hawaiian species, and this is only infrequently overcome under laboratory conditions. The overt behavioral elements of sexual isolation between Hawaiian flies are partly understood as a result of the studies of Spieth (1966, 1968, this volume). In addition to qualitative and quantitative differences in courtship routines between species, he has found differences in the field lek behavior, which are highly significant to overall sexual isolation under field conditions. Mechanical isolation is not considered a significant factor in the picture-winged group, since the genitalia of the species show little morphological differentiation (Takada 1966; Kaneshiro 1969). In at least five instances in Yang and Wheeler's study, females were found to have been inseminated, yet failed to produce progeny. The barrier here was postmating but prezygotic. Possibly, it involved the insemination reaction which is prevalent in the genus *Drosophila* (Patterson and Stone 1952).

The results from the successful laboratory hybridizations were very varied, indicating an array of postmating mechanisms between species. In some cases, the hybrids formed were of reduced viability, and died in larval, pupal or early adult stages. In other cases, hybrid viability was normal, but fertility was reduced.

Backcrossing tests indicated a range of degrees of sterility in both male and female hybrids of different combinations. Most frequently, the F_1 hybrid females were fertile, whilst the F_1 hybrid males were sterile. Amongst Hawaiian *Drosophila*, the sterility of F_1 hybrids is almost certainly always genic in nature, rather than chromosomal. Some of the species hybridized are differentiated by multiple fixed inversion differences, but most differ by only one or two inversions. Significantly, several homosequential species pairs, with identical chromosomal sequences (Carson, Clayton and Stalker 1967) are known to give sterile male hybrids (Yang and Wheeler 1969; Craddock in ms.; Kaneshiro, unpublished observations).

The kind of reproductive barrier specified by the hybridization test is not always a good indication of the degree of relationship between the species concerned. The following examples, taken from the data of Yang and Wheeler (1969), demonstrate this clearly. Production of viable, but sterile hybrids is possible in crosses between the homosequential species *D. bostrycha* and *D. disjuncta,* as well as between the distantly related species *D. hawaiiensis* and *D. discreta* which belong to different cytological subgroups within the picture-winged group. Hybrid inviability has similarly been found between closely and distantly related species. Crosses between *balioptera* females and *villosipedis* males resulted in F_1 hybrids which did not survive beyond the pupal stage, although the two parental species both belong to the *grimshawi* subgroup. *D. pilimana* females (*pilimana* subgroup) crossed with *villosipedis* males (*grimshawi* subgroup) gave hybrids which died in the larval stages. In many other instances, no hybrids at all were produced, suggesting the presence of a strong premating barrier. This was the result in reciprocal crosses between *D. balioptera* and *D. engyochracea,* two species which are very closely related in an almost direct ancestor-derivative relationship (Carson 1970). Failure to produce hybrids was even more frequent in crosses between species from the different cytological subgroups (*cf.* Table 2 of Yang and Wheeler 1969). Some of these combinations of species are separated by postmating barriers, in addition to the premating barriers. Hybrids between species pairs from different cytological subgroups have been produced artificially via the technique of ovarian transplantation (Kambysellis 1970), and such hybrids invariably show some breakdown. In five combinations, the resulting hybrids showed severely reduced viability, dying as embryos or at somewhat later stages in development. In two other instances, in crosses between the *planitibia* and *grimshawi* subgroups (*hemipeza* ♀ x *grimshawi* ♂ and *silvestris* ♀ x *grimshawi* ♂), hybrid adults were obtained, but the males showed significant sterility (Kambysellis pers. comm.).

In summary, the data indicate that premating and postmating barriers of various kinds and strengths occur between species pairs of all degrees of relationship. Some apparently closely related species, as judged by the chromosomal phylogeny, are separated by a degree of reproductive isolation equivalent to that found between distantly related species. More specifically, there seems no obvious correlation between degree of relationship and level of *postmating* isolation, as judged by extent of hybrid inviability and sterility. However, in Yang and

Wheeler's study, there were more 'successful' hybridizations between species from the same cytological subgroup than between species from different subgroups. Further, the data of Kambysellis (1970) support the expectation that the more distant species are more strongly isolated by a combination of premating and postmating mechanisms.

III THE EVOLUTION OF REPRODUCTIVE BARRIERS BETWEEN PICTURE-WINGED SPECIES

The reproductive barriers between related species are typically complex combinations of several mechanisms which could hardly be acquired instantaneously. If such barriers do indeed evolve, then it might be supposed that the reproductive isolation between recently evolved or 'new' species pairs would be much more simply based than that between evolutionarily 'old' species. In the former case, the barrier may depend on one or a small number of factors, in contrast to the complex of factors usually constituting the reproductive barrier between long established species. If these assumptions are correct, it should be possible to determine the *primary* barrier to gene flow, and then to infer the sequence of events in the development of the actual species gap. Comparison of the reproductive relationships between the most closely related and more distant species in a particular group, as well as between forms at subspecific levels of divergence where available, may provide some information on the evolution of reproductive isolation. This follows only if it can be assumed that level of phyletic divergence bears some relationship to time since differentiation.

Preliminary comparison of the available hybridization results from closely related species pairs and from distantly related species pairs initially does not appear to indicate any particularly strong trends in the evolution of reproductive isolation in Hawaiian *Drosophila,* since the kind and extent of isolation at each level of relationship is extremely variable. It could be that possible correlations are obscured by the broad comparative approach. It could also be that the distantly related species being compared are too far removed, so that interpretation of their reproductive isolation is biased by the intervention of partially dependent but irrelevant elements. In any case, the more pertinent comparison for our purposes is not between closely and distantly related species, but rather between the most closely related species pairs at times variously removed from their initial divergence. Investigation of direct interrelationships between species on a time scale is more likely to provide information on the critical early stages in the development and reinforcement of reproductive isolation than investigation of interrelationships on a phyletic scale. Therefore, in the remainder of this discussion, particular attention will be paid to the most closely related and most recently differentiated species, with the intention of defining the primary barriers which are operating to curtail gene flow. Species will be judged to be very closely related if they are only one step removed from one another with no known intermediate forms, in the cytological phylogenetic scheme of Carson (*cf.* Clayton, Carson & Sato 1972 for updated version of the relationship chart). Usually such directly related species pairs differ by only one or a small number of inversions. In a number of cases, they are homosequential.

The time since differentiation of particular species pairs can never be known accurately, but rough estimates of comparative species ages can be made on the basis of the available geological information. In this context, species from the Big Island of Hawaii will be considered evolutionarily young, and species from the island of Kauai, the most northerly of the windward islands, will be considered evolutionarily old (although still very young by most standards). According to the potassium-argon datings of McDougall (1964), the entire Big Island of Hawaii is less than one million years old. All of the members of the picture-winged group found here are unique to this island, and must have differentiated on this island. These species are undoubtedly the most recent evolutionary products of the picture-winged group. Kauai is estimated to be 5.6-3.8 million years old, and is presumed to be the primary site of origin of present members of the picture-winged group, on the basis of both geological and chromosomal evidence (Carson *et al.* 1970). Accordingly, Kauai species are considered to be the most primitive and most ancient representatives of the group. The relative antiquity of the present-day species is admittedly open to question. Although some of them might be relicts of the original primitive species presumed to have arisen shortly after the first colonization of the Hawaiian Islands by *Drosophila,* others could well be more recent speciation products. It is even possible that some could be of equivalent age to the 'young' Hawaii species. In view of this element of doubt, greater emphasis will be placed on the characteristics of the obligatorily recent Hawaii species.

It is important to recognize that there may be more than one mode of speciation involved in the evolutionary diversification of this group of *Drosophila,* and more than one sequence of events in the evolution of reproductive isolation. Different situations should therefore be assessed separately. The geographic relationships of species provide one obvious basis for classification, and these will be used to distinguish different situations in the ensuing discussion.

The case of allopatric species

Many of the sets of chromosomally close, and therefore phyletically closely related species in the picture-winged group have representatives on each of the islands. Thus many species are geographically isolated from their closest relative by an expanse of ocean which very effectively prevents gene flow in either direction. The question of the reproductive isolation or lack of it between such geographically isolated or allopatric species is rather an academic one, since under normal circumstances, such species would never have an opportunity to hybridize and exchange genes, and the validity of their reproductive isolation would never be challenged. However, if their allopatric separation involved a *bona fide* speciation event, there should be a measure of reproductive isolation between them. This is on the assumption that the Hawaiian forms recognized morphologically and behaviorally as species fit the generally accepted species criteria. Since, in the Hawaiian situation, the allopatric relationships of species are not all of exactly the same kind, their reproductive relationships may not be uniform. The following three categories will therefore be discussed separately:

    (*a*) allopatric species distributed between the major islands;

    (*b*) allopatric species distributed between the islands of the Maui-complex;

    (*c*) allopatric subspecies.

(*a*) *Allopatric species distributed between the major islands*

The major islands of the Hawaiian group inhabited by *Drosophila*, in sequence from the north-west to the south-east are Kauai, Oahu, Maui and Hawaii (Figure 1). Immediate allopatric derivatives of the Kauai species are

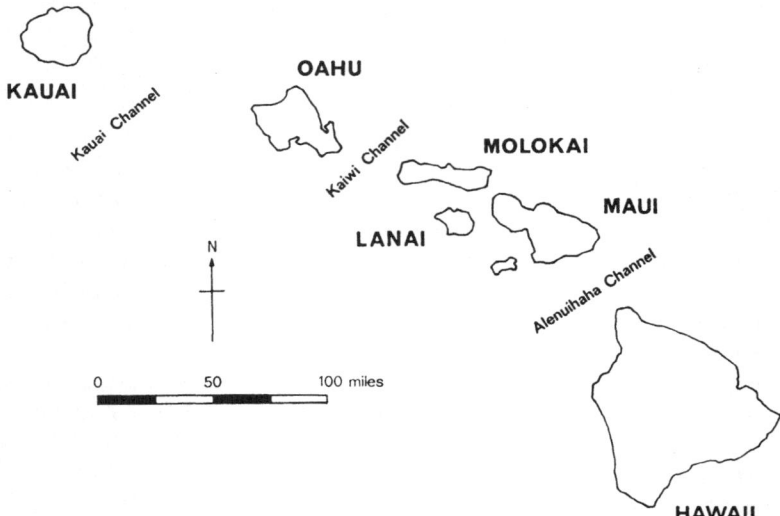

Figure 1: The windward islands of the Hawaiian Archipelago, which support endemic *Drosophila* species.

found on each of the other islands. Data on the reproductive relationships of such species are very limited, since few of the Kauai species have been appropriately tested in hybridizations. Table 1 presents the available information for two Kauai species. *D. villosipedis* has been hybridized with five closely related allopatric species, four of which are homosequential with *villosipedis*. *D. crucigera* has populations on both Kauai and Oahu, and many of Yang and Wheeler's inter-specific hybridizations with this species used the Oahu strain or a mixture of Oahu and Kauai strains. Since there is possibly some differentiation between these two *crucigera* populations, (*cf*. later discussion in (*c*)), only one of Yang and Wheeler's results can be included in Table 1. The failure of hybridization in most instances suggests the general presence of premating barriers between Kauai species and their allopatric relatives. Where this barrier is absent, or overcome under laboratory conditions, postmating barriers of relatively severe hybrid inviability and sterility are revealed. Although generalizations are premature in the face of so little data, the reproductive isolation of the 'old' Kauai species

Table 1. Reproductive relationships between two presumably 'old' Kauai *Drosophila* species (*D. villosipedis* and *D. crucigera*) and their allopatric close relatives. Data selected from the hybridization results of Yang and Wheeler (1969).

Interspecific hybridization Female Male	Chrom.† Rel'ship	Premating barrier	Postmating barrier	
			Hybrid inviability	Hybrid sterility
villosipedis x *disjuncta*	H	Yes		
disjuncta x *villosipedis*	H	Yes		
bostrycha x *villosipedis*	H	Yes		
villosipedis x *orphnopeza*	H	Yes		
villosipedis x *grimshawi*	H	Yes		
grimshawi x *villosipedis*	H	—	Most larvae died in 2nd or 3rd instar: ♂♂ died before maturing	Surviving ♀♀ sterile
villosipedis x *pilimana*	1 invers.	Yes		
pilimana x *villosipedis*	1 invers.	—	Few nonviable larvae died in 2nd instar	
disjuncta x *crucigera*	1 invers.	—	Larval death in early instars	

† H—The two species hybridized are homosequential. The nonhomosequential pairs are differentiated by one fixed inversion.

Table 2. Reproductive relationships between 'recent' *Drosophila* species from the Island of Hawaii and their allopatric close relatives.

Interspecific hybridization Female	Male	Chrom.† Rel'ship	Premating barrier	Postmating barrier		Source φ of data
				Hybrid inviability	Hybrid sterility	
*paucipuncta x	uniseriata	H	—	—	♂♂ sterile, ♀♀ fertile	Y & W
uniseriata x	paucipuncta	H	—	—	♂♂ sterile, ♀♀ fertile	Y & W
*silvestris x	planitibia	H	—	—	♂♂ sterile, ♀♀ fertile	EMC
planitibia x	silvestris	H	—	—	♂♂ sterile, ♀♀ fertile	EMC
*heteroneura x	planitibia	H	—	—	♂♂ sterile, ♀♀ fertile	EMC
planitibia x	heteroneura	H	—	—	♂♂ sterile, ♀♀ fertile	EMC
*silvarentis x	gymnobasis	H	—	—	F_1 ♂♂ sterile	KYK
gymnobasis x	silvarentis	H	—	—	F_1 ♂♂ sterile	KYK
*hawaiiensis x	gymnobasis	H	—	—	F_1 ♂♂ sterile	KYK
gymnobasis x	hawaiiensis	H	—	—	F_1 ♂♂ sterile	KYK
*hawaiiensis x	gradata	2 invers.	—	—	F_1 ♂♂ & ♀♀ semisterile	Y & W
gradata x	hawaiiensis	2 invers.	—	—	not tested	Y & W
*formella x	'villitibia' (Maui complex)	polymorphic for 2 new inversions				
'villitibia' x	formella		—	—	F_1 ♂♂ sterile	H & K
grimshawi x	*pullipes	H	?	—	F_1 ♂♂ sterile	H & K
*ochracea x	grimshawi	1 invers.	?	F_1 ♀♀ viable, ♂♂ inviable (died at 2nd larval instar)	F_1 ♂♂ sterile	H & K
*sproati x	grimshawi	1 invers.	?	inviable — died at 3rd larval instar	?	MPK
'engyochracea x	balioptera	1 invers.	Yes	?	?	Y & W
balioptera x	engyochracea	1 invers.	Yes	?	?	Y & W
*adiastola x	*setosimentum	10 invers.	—		F_1 ♂ sterile	HLC

* The Hawaii species. The other parent in the hybridization is from one of the other islands.

† H—The two species hybridized are homosequential. Nonhomosequential pairs are differentiated by the number of fixed inversions indicated.

φ The data are taken from the following authors

Y & W—Yang and Wheeler (1969).

EMC—Craddock (in ms.).

KYK—Kaneshiro (unpublished observations).

H & K—Hardy and Kaneshiro (1972).

MPK—Kambysellis (1972).

HLC—Carson (unpublished observations).

seems to depend most often on premating barriers, and less frequently on well developed and highly effective postmating barriers.

Hybridization data involving the recent Hawaii species are more abundant, and the known reproductive relationships of Hawaii species with closely related allopatric species, from Maui and Oahu, are presented in Table 2. Strong sexual isolation, or other kinds of premating barrier seem to be absent in all cases except one. The laboratory hybrids produced are mostly viable, with the exception of two species combinations which showed larval inviability. The evidence indicates that, between recently differentiated pairs of allopatric species, the predominant barrier to free genetic exchange is posed by the sterility of male hybrids. Female hybrids were found to be fertile whenever tested, in accordance with the pattern frequently found between allopatric continental *Drosophila* species (Patterson and Stone 1952). Whereas there is usually a breakdown in the fertility of the resulting backcross progeny, in the case of some of the Hawaiian species at least, this is not so. Only hybrids of the *planitibia-silvestris* and *planitibia-heteroneura* combinations have been tested in a second generation of back-crossing, but in both these cases, male and female backcross progeny derived from the F_1 hybrid female proved to be fertile (Craddock in ms.). These situations might offer some potentiality for gene exchange, were it not for the presence of an absolute geographic barrier between the two species. This physical ocean barrier provides a means of premating isolation. However, genetic isolation between these and most of the other recent species depends on partial postmating barriers resulting from the sterility of F_1 male hybrids.

Comparison of the forms of reproductive isolation between old and recent allopatric species (Tables 1 and 2) suggests that the primary barrier between closely related species distributed on the different major islands most often operates via male hybrid sterility. Hybrid inviability, and premating or sexual isolation appear to be secondarily developed, so that older pairs of allopatric species show a much greater degree of reproductive isolation than more recently differentiated species pairs. Some of the Hawaii species show only minimal isolation from their presumed allopatric ancestor, together with a measure of morphological differentiation, sufficient to warrant their initial recognition as distinct species.

(b) Allopatric species distributed between the islands of the Maui-complex

The case of the forms and species distributed on the different islands of the Maui-complex has certain special features which make it worth considering as a separate category. The complex is made up of four islands (Figure 2): three of these, Maui, Molokai and Lanai, currently support *Drosophila* populations. There is strong evidence that Pleistocene land bridges once linked all islands of the complex (Stearns 1966), and it is probable that *Drosophila* populations were once co-extensive across the lower-lying regions connecting the separate volcanoes. Furthermore, the islands have probably been joined and separated from one another at least twice in the recent geological past, as a consequence of rises and falls in sea level. The present geographic separation of the islands

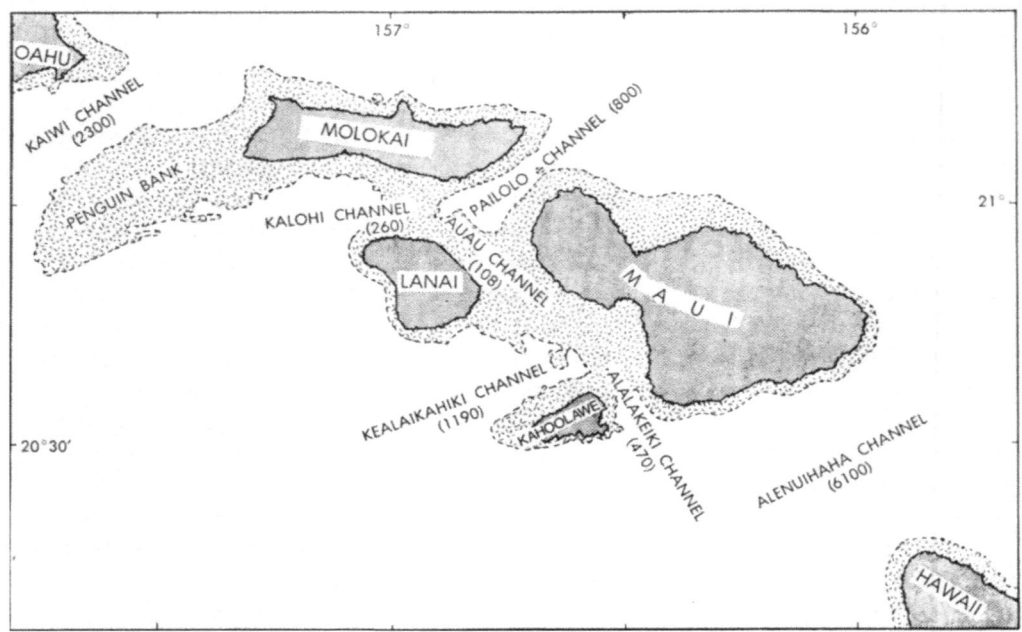

Figure 2: The Maui-complex of islands, Maui, Molokai, Lanai and Kahoolawe. The separate volcanoes have been joined and separated from one another at least twice during the Pleistocene, as a result of rises and falls in sea level. The channels between the islands are comparatively shallow, by contrast with the deep ocean channels between the other major islands.

means that many species of the Maui-complex now consist of three or more allopatric populations. The largest island of the complex, Maui, consists of two volcanic domes, East Maui and West Maui, which arose independently. Although now physically connected via a narrow low-lying saddle, these two volcanoes are currently biologically quite separate. The island of Molokai is also composed of two shield volcanoes, East Molokai and West Molokai. Effectively therefore, the Maui species probably have one population per volcano. In most cases, the reproductive unity of these species has never been adequately tested. It has usually been proposed that the Maui and Molokai populations belong to the same species, since these islands have been connected comparatively recently. However, between some allopatric populations on different islands and volcanoes of the complex, there is a measure of reproductive isolation, which, in a few instances, has been equated with specific differentiation.

The species *bostrycha* and *disjuncta*, which are endemic to Molokai and Maui respectively, are very closely related: they are homosequential and they share the same inversion polymorphism in their populations. Reciprocal crosses between these two species produce viable hybrids; the female progeny are fertile, but the males are sterile (Yang and Wheeler 1969). The species are extremely similar morphologically, differing solely by two tibial hairs.

Lack of strong morphological differentiation between certain closely related allopatric forms of Hawaiian *Drosophila* may often have prevented recognition of distinct species. In fact, there are several examples now requiring taxonomic revision, because inter-island hybridization tests have demonstrated the existence of a reproductive barrier, in addition to the geographic barrier. Closer examination has sometimes revealed very minor morphological differences between the different island populations, correlated with their reproductive differentiation. Some of the examples of Maui-complex populations currently under review are cited below.

The species name *villitibia* originally referred to all the Maui-complex populations of a particular *hawaiiensis*-group fly. It has since been found that Maui and Molokai populations are not interfertile (Kaneshiro pers. comm.). In crosses between '*villitibia*-like' females from East Maui and *villitibia* wild caught males from Molokai (the type locality), some of the F_1 progeny died in the late third instar, and in pupal stages; amongst surviving adults there was an unequal sex ratio with a large excess of adult females; and the testes of the adult males produced were underdeveloped and lacked sperm. The reproductive isolation between these two forms thus involves postmating effects dependent on a degree of hybrid inviability and hybrid sterility. Morphologically, the differences between them are only very slight, being based on a difference in their wing pattern, and a difference of three or four hairs in the ciliation on the tibia. The two populations differ further in their metaphase karyotypes (Clayton 1968, and unpublished observations), and together these differences are sufficient to warrant species recognition. The '*villitibia*-like' form from Maui is currently being described.

D. planitibia is another form which occurs on more than one island of the Maui-complex. Crosses between populations from three volcanoes, East Maui, West Maui and Molokai, produce viable F_1 hybrid adults, which show some reduction in fertility (Craddock in ms.). It has been found that F_1 male hybrids of the East Maui-Molokai combination are sterile in backcrosses, due to irregularities in sperm maturation. The F_1 hybrid females were fertile. The fertility barrier between these two populations of *planitibia* is correlated with very slight morphological differentiation—a coloration difference has recently been discovered by Mr. K. Kaneshiro—and with striking allelic differentiation (Johnson, pers. comm.). The three populations appear to be genetically distinct, and probably each is reproductively isolated from the other two by postmating barriers of partial hybrid sterility.

The species described as *D. odontophallus* is distributed on Molokai, and East and West Maui, and flies from the three populations are morphologically identical, as far as is known, and also chromosomally identical. Inter-island hybrid males from crosses between Molokai females and West Maui males were sterile in backcrosses to both parental populations. Most of the hybrid males completely lacked motile sperm but some contained a few sperm with very much reduced motility (McNutt and Carson, unpublished observations). Again, the partial reproductive isolation indicated by hybridization tests is correlated with significant genetic differentiation between the populations concerned (Johnson, pers. comm.). The total divergence in this case seems to be less than in others, since morphological differences are apparently lacking, and hybrid males are perhaps only semisterile.

The cases discussed demonstrate a pattern of allopatric population different- iation and evolving reproductive isolation at the postmating level. Typically, related populations of the Maui-complex show little or no obvious morphological differentiation; in some cases, a significant degree of genetic differentiation in isozyme frequencies has been detected (Rockwood *et al.* 1971; Johnson, in ms.). Presumably, the East Maui, West Maui and Molokai populations of different species were joined and finally isolated geographically from one another at much the same time, so that comparisons of the degrees of reproductive isolation be- tween recently and long separated populations are not possible in this instance. What can be stated is that there is probably considerable variation in the rate of evolution of reproductive isolation between the Maui-complex populations, if indeed it originates at all. Some species show no significant interpopulation variation in isozyme frequencies (Johnson, in ms.), and when tested, might be found to be interfertile between the various Maui populations. Until this is done, each allopatric population should be considered a separate reproductive com- munity, which is potentially isolated reproductively from the populations with which it was formerly in geographic and genetic contact.

(c) Allopatric subspecies

In this third category of allopatric forms, I place the few species of the picture-winged group which have a distribution encompassing more than a single major island. Multi-island distribution is very atypical in the group, and is probably confined to just two species—*D. crucigera* and *D. grimshawi*. The allopatric populations of these two species supposedly do not show any repro- ductive isolation. Other data, however, indicate some population differentiation within each, which might perhaps be preparatory to the evolution of a reproduc- tive difference. Currently, the various island populations could be referred to as geographic subspecies.

D. crucigera occurs on Oahu and Kauai, and flies from the two islands are morphologically identical (Hardy 1965). Populations from the Kokee-Halemanu area of Kauai to the east and from windward Kauai to the west are chromo- somally distinct, as are populations of *crucigera* from the Waianae Range and from the Koolau Range of Oahu (Carson 1966; Carson *et al.* 1970). These populations differ in their frequencies of polymorphic gene arrangements on the X-chromosome and on the third chromosome, and have been called chromosomal races. All available crossing data suggest that the populations are completely interfertile, and therefore members of the same biological species. However, Yang and Wheeler (1969) note that the Oahu and Kauai forms behave differently in crosses to other species. It is therefore conceivable that the different geographic forms are genetically differentiated to some extent, but significantly less so than is usual between forms distributed between the major islands (*cf.* the allopatric species considered in section (*a*)).

D. grimshawi has been collected on Kauai, Oahu, Maui, Molokai and Lanai. There is as yet no evidence of reproductive isolation between any of these populations (Carson *et al.* 1970; Carson 1971), although only some of the inter-island hybridizations have been carried out. Populations differ substantially in the frequency of a polymorphic chromosome 4 inversion sequence, even within an island (Carson and Sato 1969), but all have been regarded as conspecific. Nonetheless, there are significant differences between the various insular forms. Kauai and Oahu *grimshawi* differ morphologically from flies from the Maui-complex of islands, as well as differing in their relative abundance and ease of rearing in the laboratory. Further ecological differences between populations have recently been found, involving polyphagous vs. monophagous host preferences (Carson, pers. comm.). Oahu *grimshawi*, moreover, are differentiated from the other insular subspecies of *grimshawi* by a very slight chromosomal difference. In addition, *grimshawi* flies from the various islands behave differently from one another in crosses to other species (Yang and Wheeler 1969). All of these factors suggest some genetic differentiation between the insular subspecies of the complex, which might perhaps be correlated with a barely minimal degree of reproductive isolation between some of the populations.

On the Big Island of Hawaii, there is a form, extremely closely related to *D. grimshawi* but recognized as a separate species, *D. pullipes* (Hardy and Kaneshiro 1972). The Big Island population is chromosomally identical with the Oahu population of *grimshawi*, and morphologically almost identical with the other populations, differing from the Maui *grimshawi* only in the colour of their legs. Hybrid males from crosses between Maui *grimshawi* females and Hawaii *pullipes* males are sterile (*cf.* Table 2). It is of further interest that the Oahu *grimshawi* and *pullipes* of Hawaii possess the same host relationship. In view of the partial reproductive isolation of the Hawaii population from other members of the complex, it is quite likely that previously undetected isolating barriers might also exist between other populations. Further studies of these forms are in progress.

The evolution of reproductive isolation between the insular populations of *D. crucigera,* and of *D. grimshawi* would appear to be a secondary event, which may or may not follow after a certain amount of genetic differentiation has taken place. The genetic divergence might involve aspects other than morphology, and might be related to total adaptation. Unfortunately, there are no data available on the isozyme variation between the various insular forms of these two species. These would be valuable for comparison and assessment of the genetic events occurring in populations which might be extremely close to speciation. Reproductive barriers in this situation would depend initially on postmating isolation via sterility of the male hybrids.

The level of insular differentiation in *D. grimshawi* and *D. crucigera* contrasts strongly with the normal pattern within the picture-winged group, and indeed within the Hawaiian *Drosophila* as a whole. The typical situation is that considered earlier under (*a*), in which closely related allopatric forms distributed on the different major islands are markedly distinct morphologically, isolated from one another by substantial reproductive barriers, and very clearly separate species, each endemic to its own island. The specific or subspecific differentiation in the *crucigera* and *grimshawi* complexes is of a rather different kind. The Big Island species, *D. pullipes,* is only marginally distinct from the species '*D. grimshawi*'. Although unusual, this pattern is not unique; there are at least two parallel cases which involve a series of very closely related allopatric populations on the various islands, one member of which is reproductively isolated from the others. This reproductively isolated form is, in each case, morphologically identical with the other members of the complex, or almost so, and as such, had been equated with them as the same species, until the inter-island hybridization tests had been performed. One example is that of *D. liophallus,* originally considered to be distributed on Oahu, East and West Maui and Molokai. Hybridization between the Oahu and West Maui populations produced sterile male hybrids which showed extensive cellular degeneration of the testis (McNutt and Carson, unpublished observations), establishing the presence of a reproductive barrier between these two morphologically identical populations. The Oahu population has now been recognized as a separate species, *D. gymnophallus,* distinct from the Maui-complex populations of *D. liophallus.* The other example is that of the Hawaii population of the '*villitibia*' complex, which has now been described as *D. formella* (Hardy and Kaneshiro 1972). This species is only very slightly different morphologically from the Maui form, but it is chromosomally distinct in its mitotic metaphase karyotype and in the presence of two new polymorphic inversions (Clayton *et al.* 1972). This group of forms has the added complication of reproductive differentiation between allopatric populations of the Maui complex of islands (*cf.* section (*b*)). Cases such as this and such as the *grimshawi* complex, demonstrate more than one pattern of allopatric differentiation and evolving reproductive isolation even within such a close set of populations.

The case of sympatric species

A consideration of the degrees of reproductive isolation between sympatric species is biologically more meaningful than an assessment of isolation between allopatric species, since populations in contact must evolve effective isolating mechanisms to retain their identity. In a sympatric situation, selective pressures favouring build-up of such mechanisms would be much stronger than in an allopatric situation, and one might expect more specific and more complete forms of isolation to have developed. Also, it might be expected that the basis of the isolating mechanisms might differ from that for allopatric species.

Most Hawaiian *Drosophila* habitats support a considerable number of *Drosophila* species, some of which are very closely related. The reproductive isolation of these species is constantly subject to test, whether their derivation was recent or ancient. There is no evidence whatsoever for natural hybridization in Hawaiian *Drosophila,* which suggests that sympatric species are isolated by strong premating barriers which are highly effective under field conditions. The barriers often persist under laboratory conditions, since many of the crosses attempted between sympatric species failed to produce any F_1 hybrids (Yang and Wheeler 1969). Where successful, the crosses reported by Yang and Wheeler invariably produced F_1 hybrid males that were sterile. Sometimes, females from such hybridizations were also sterile. There were also some instances of hybrid inviability between sympatric species, e.g. the F_1 hybrids from both reciprocal crosses between the Maui species *balioptera* and *disjuncta* died in the pupal stages. These species are comparatively closely related. In general then, the evidence supports the existence of strong postmating barriers between sympatric species (dependent on either hybrid inviability or hybrid sterility), in addition to the universal occurrence of strong premating barriers.

Data on hybridizations between sympatric species which are the most closely related representatives of a particular grouping are very limited. No comparisons are available for the old Kauai species: the few data available for recently derived pairs of sympatric species (Big Island) are presented in Table 3. For both combinations of species, the premating isolation is overcome in only one direction, permitting partial assessment of the postmating barriers. The *hawaiiensis-silvarentis* pair of species are separated from one another by a postmating barrier dependent on hybrid sterility, as well as by the premating barrier. Where there are multiple barriers such as this, it is impossible to say which one originated first. The *silvestris-heteroneura* combination however, appears to have but a single barrier to genetic exchange—the premating barrier which is the operative mechanism in the field. We can therefore state, with confidence, that implementation of a premating barrier has been the primary event in the evolution of reproductive isolation between this pair of sympatric species.

The lack of a strong sterility barrier between *D. silvestris* and *D. heteroneura* is very surprising. There is no evidence of any hybrid breakdown, both sexes of F_1 hybrids, backcross hybrids etc. being completely fertile. The hybridization data and other evidence are presented in detail elsewhere, together with a fuller discussion of this situation (Craddock, in ms.). Isolation between the two species might be judged to be rather precarious, but it is apparently effectively maintained under field conditions, and the two can be regarded as valid species. Morphologically, they are quite distinct: chromosomally and biochemically however, they are extremely similar. The nature of the premating barrier between them has not yet been defined. The two species utilize the same host plant, and do not seem to show any ecological separation. Further, the lekking sites chosen by males appear to be identical, so there is little, if any microspatial separation between them. It is suggested that their reproductive isolation depends on subtle behavioral factors, perhaps related to the sensory components of mating behavior. Their overall mating behavior patterns are very similar, except for a minor difference in the preliminary segment of the courtship when the female is made aware of the presence of a courting male (Spieth, unpublished ms.). This difference, of itself, would seem not to be sufficient to prevent interspecific copulation in the field, but together with possible differences in stimuli, it could contribute to the completion of sexual isolation between the two species. The

Table 3. Reproductive relationships between 'recent' Hawaii species and their closest sympatric relative.

Interspecific hybridization Female Male	Premating barrier	Postmating barrier		Source of data
		Hybrid inviability	Hybrid sterility	
heteromeura x *silvestris*	strong sexual isolation (F & L) *	?	?	EMC
silvestris x *heteromeura*	sexual isolation (F)	none	none	EMC
silvarentis x *hawaiiensis*	strong sexual isolation (F & L)	?	?	Y & W
hawaiiensis x *silvarentis*	sexual isolation (F)	—	♂ ♂ sterile	Y & W

Both pairs of species are homosequential.

* F & L—Isolation is complete in the field and in the laboratory; F—Isolation is effective in the field but is broken down under laboratory conditions.

mating barrier is significantly stronger in one direction than the other. Males of *silvestris* consistently failed to achieve insemination of *heteroneura* females, although the reciprocal cross was successful in a considerable proportion of cases. The circumstances which could lead to the sympatric establishment of such a behavioral mating barrier, in the absence of other kinds of differentiation, are hard to envisage, yet there is no compelling evidence that the species were ever allopatric.

Sexual isolation is extremely well developed in Hawaiian *Drosophila*, and is probably the most important barrier between sympatric species. Many species have highly intricate courtship routines, and secondary sexual characteristics are extensively developed (Spieth 1966, 1968). In the field, the lek behaviour of males probably contributes as much to interspecific isolation as the differences in courtship behavior patterns.

The premating barriers between sympatric species of Hawaiian *Drosophila* might also involve ecological differentiation between the species concerned. Utilization of different hosts by two related sympatric species is known in a few instances, and this factor could well help to separate them at the premating level. The homosequential species, *D. silvarentis* and *D. heedi* from the Big Island, occur in mixed populations and appear to be ecologically isolated via their adaptation to feeding and oviposition in different sites of the slime fluxes associated with *Myoporum* trees (Kaneshiro *et al.* 1973). It is not known whether this is the sole factor in their premating isolation, or whether sexual behavior differences also play some role. Further, there is no hybridization data available to assess the presence and extent of postmating barriers between them. Unfortunately, evidence of this kind is lacking for most sympatric species pairs. At this stage, it is difficult to estimate how often the sexual isolation of sympatric species pairs is associated with well developed postmating barriers, and how often ethological or other premating barriers are the sole determinant of interspecific isolation, as in the case of *D. silvestris* and *D. heteroneura*.

IV PATTERNS OF SPECIATION IN HAWAIIAN FLIES

Dobzhansky (1972) recently concluded that 'there is not a single kind but there are several kinds of species and of processes of speciation in *Drosophila*'. This is nowhere so evident as amongst the Hawaiian *Drosophila*. Furthermore, because of the special nature of the insular situation, insular speciation patterns may demonstrate additional features, or entirely different ones from those involved in the continental situation. The reproductive relationships between Hawaiian species are highly variable, indicating a great diversity in the speciation mechanisms involved. Species barriers may have been formed in several different ways in the range of circumstances which exist in island populations. Comparison of the nature of these reproductive barriers, and of the degrees of isolation which they afford, might indicate the mode of speciation associated with each kind of situation. One complication arises in that the pattern of species relationships for a particular situation is not always uniform. This variation probably results from the varying ages of different species sets, and the varying rates of species differentiation associated with different ecological conditions at various times and under various circumstances. In spite of these difficulties, some tentative hypotheses can be proposed as to the speciation modes involved in the evolutionary radiation of the Hawaiian *Drosophila*.

Geographic isolation has been of primary importance in the initiation of species divergence in the Hawaiian Islands. It has operated in more than one way, and at both the inter-island and the intra-island or local levels. However, geographic barriers might not have been the only factors responsible for initial separation of gene pools. This possibility should be kept in mind in considering the probable speciation patterns in this group.

The allopatric species

The differentiation of all the allopatric *Drosophila* species has obviously not occurred via the same route. Each of the three distribution patterns considered suggests a different speciation pattern and a different genetic sequence in the origin of reproductive isolation. The patterns associated with each of the categories can be characterized as follows:

(*a*) founder event after inter-island dispersal

(*b*) population splitting of one originally continuous population

(*c*) insular speciation consequent upon inter-island migration, but not directly determined by founder effects.

(*a*) The predominant speciation pattern in the Hawaiian situation is that associated with an inter-island founder event. With few exceptions, the different major island population sets which are most closely related chromosomally are specifically distinct, with a different species on each of the major islands. In other words, if the founder individual was successful, almost invariably a new species resulted. Allopatric species originating in this manner are subject to an extreme form of geographic isolation. The oceanic channels between the islands represent a major geographic barrier to *Drosophila*. Migrations between the islands are rare, successful colonization and establishment are even less probable, and effective genetic contact between the established founder population and its ancestral population, via inter-island migration, must be minimal, if not completely absent. If the colonization is effected by a single fertilized female, as is probably the usual case, the new founder population is likely to undergo severe inbreeding and a 'genetic revolution' (Mayr 1954).

A scheme of the population events which might follow a founder event and lead to speciation has been proposed by Carson (1971). In his view, reproductive isolation originates relatively rapidly after the founder event, in only a small number of generations, as a result of the non-adaptive genetic changes which occur in the founder population. The data reviewed here suggest that allopatric species of picture-winged *Drosophila* with a 'founder' origin are isolated primarily by a postmating barrier, dependent initially on male hybrid sterility. This genetic barrier is only a marginal one which would not be effective were it not for the presence of the geographic barrier. Substantial gene flow between 'species' would still be possible as a consequence of the fertility of hybrid females. However, with time, the primary reproductive barrier between such allopatric *Drosophila* species apparently evolves to include hybrid inviability and perhaps premating factors. The strengthening of reproductive isolation between the older allopatric species might be correlated with the increasing genetic divergence between the ancestral and the derived populations which would result as the new species entered an adaptive phase. The observations concur with expectations such as these which can be drawn from Carson's hypothesis of speciation by a founder event. Reproductive isolation in these circumstances is thought to arise prior to adaptive genetic differentiation, rather than as a consequence of it, as in traditional interpretations of geographic speciation (*cf.* Mayr 1963; Dobzhansky 1970). Evolution of a genetic barrier between populations differentiated by a founder event, and therefore 'founder-type speciation', would thus occur completely by chance. Origin of the primary barrier would be non-adaptive, as would be subsequent reinforcement of this isolation. Such reinforcement actually results from adaptive processes *within* the new species, but cannot be viewed as directly adaptive in the sense of promoting isolation of the new species from its allopatric ancestor.

(*b*) The second kind of allopatric species formation represented by species and 'subspecies' populations of the Maui-complex also relies on geographic isolation, but in a different way. It is inferred that population distribution of particular species over the Maui-complex was more or less continuous at some time in the past. Isolation between the populations of the different volcanoes has been imposed secondarily. Populations must have been divided as a result of rises in sea level and isolation of the individual islands by ocean barriers, and as a result of intra-island changes in the habitat and vegetation at lower altitudes between volcanoes which accompanied the climatic changes. This process would not have involved the close inbreeding which necessarily follows a founder event. Further, the possible secondary contacts between the different island isolates at the times of land bridge connections would tend to maintain a degree of genetic continuity. Many of the Maui populations show only minor genetic differentiation and are merely races or subspecies with no evidence of reproductive differences between them. Some however, are partially isolated by postmating barriers involving male hybrid sterility and occasionally hybrid inviability. No cases of premating isolation are known, although perhaps this may develop if the geographic barriers are maintained.

The speciation pattern of the Maui populations is analogous to the more usual form of geographic speciation (Mayr 1963). In the presence of a geographic barrier, populations undergo a primary phase of adaptive genetic differentiation, which may or may not eventually lead to the *secondary* development of reproductive isolation. In some cases this may have evolved selectively at times of secondary contact between populations. This mode of speciation contrasts strongly with that proposed for the founder situation, in which the origin of reproductive isolation is the primary event, and adaptive differentiation the secondary one (Carson 1970, 1971). As a corollary, the respective rates of speciation are widely different. In the founder situation, species formation may follow more or less immediately after the geographic separation. In the situation within the Maui-complex, species formation would occur long after the geographic separation and much more slowly, if at all. Even between the various Maui species, all subject to population splitting at much the same time, rates of speciation are highly variable.

(c) The third category concerns cases where there has been inter-island migration without 'founder-event speciation'. It applies to subspeciation within two species, and perhaps full speciation in two species pairs. Whereas, for most of the picture-winged group, inter-island migration has resulted in the formation of new species, for the two forms *D. grimshawi* and *D. crucigera,* it has not. The differentiation on Big Island of *D. pullipes* from the *grimshawi* complex is an exception. This could be considered as a final product of the differentiation which is still occurring in the other insular populations. In other words, this mode of speciation might entail a gradual process. The species gap might not occur abruptly as in the normal island founder situation; rather a series of intermediate forms or subspecies might result. Another alternative explanation is equally likely—that within the '*grimshawi* complex', there are at least two distinct speciation modes operating. *D. pullipes* on Hawaii might have been differentiated by the usual founder method of speciation. The other island populations might be following an entirely different mode. It is not even necessary that these related forms be directly linked in one sequence.

The insular subspecies of *grimshawi* and of *crucigera* might be expected to diverge via a gradual accumulation of genetic differences, which might eventually lead to a reproductive difference as in regular geographic speciation. However, speciation is by no means an inevitable end-result of this kind of genetic divergence. The present atypical lack of reproductive isolation between these island populations might be explained as a consequence of a difference in the nature of the founding event—that instead of the new population being founded by a single immigrant individual, it was founded by a comparatively large sample of individuals from the parent population. Only a very rare combination of circumstances would result in the simultaneous migration of many individuals to precisely the same area on a neighboring island. An alternative explanation is that the founder event does not always result in a genetic revolution, major genetic reorganization and the origin of reproductive isolation.

In both *grimshawi* and *crucigera*, there seems to be repetition of the same unusual pattern. Migrations of *grimshawi* to several islands have had the same consequence. Moreover, Carson argues that the two Oahu populations of *D. crucigera* were derived from two separate founder and colonization events from Kauai (Carson *et al.* 1970), neither of which apparently resulted in reproductive isolation. It may be suggested that species constitution is significant to the outcome of founder events. As yet, it is not possible to suggest the nature of the genetic factors which may be critical in this regard, or to interpret the sequence of events in the development of this pattern.

The two species pairs which may have resulted from this kind of process are the *liophallus-gymnophallus* pair and the '*villitibia*'-*formella* pair. Both are morphologically cryptic, by contrast with most allopatric pairs with a founder origin, between which morphological differentiation is well marked.

The sympatric species

The patterns of differentiation of species pairs which are now sympatric are much more difficult to deduce. There is little present basis even for deciding whether a sympatric or an allopatric origin is the more probable, or whether both may have occurred. The possibility of sympatric speciation via host plant selection has been explored by Heed (1971), and he concludes that if it does occur at all in Hawaiian *Drosophila* it is exceedingly rare. In the picture-winged group, differences in host plant specificity between closely related sympatric species are unusual.

There is a possibility of sympatric speciation via changes in sexual behavior patterns. According to Spieth, the male courtship displays in Hawaiian flies are species-specific, and mating behavior differences constitute the major isolating mechanism between many, if not most sympatric species. The occurrence of lek behavior in Hawaiian *Drosophila*, with the correlated intensification of sexual selection, has even greater potential significance than the differences in courtship patterns. In some differentiation can occur in the lekking sites chosen by males within a population, this could leave the way open for the sympatric evolution of a degree of reproductive isolation, which might then be reinforced in various ways leading to complete speciation. Spieth notes that many of the specialized structural modifications found in males must have evolved subsequent to the behavioral characteristics associated with these structures (Carson *et al.* 1970). Thus the usually well marked morphological divergence between the males of closely related sympatric species is probably a secondary phenomenon. From the data on the reproductive relationship between *D. silvestris* and *D. heteroneura* (Craddock, in ms.), it appears that a sexual behavioral difference can be the primary barrier to evolve between two species in the absence of other kinds of differentiation. This kind of premating barrier might conceivably originate sympatrically, via disruptive selection for behavioural characteristics. Of course, it could also arise allopatrically as a byproduct of geographical isolation, as favored by Spieth (1968). In the subsequent sympatric phase, positive selection pressure could well operate to further increase the initial chance divergence in mating behavior. There is no direct evidence available which differentiates between these two alternatives.

The physical characteristics of Hawaiian islands and the frequent occurrence of lava flows result in many local geographic barriers between populations on an island. *Drosophila* populations in different patches of forest (kipukas) may sometimes be isolated as effectively as if they were on different islands. The differentiation of some of the species which are now sympatric might thus have had a strong allopatric component. At least two different allopatric sequences are possible—either the chance origin of a primary premating barrier, which may or may not be followed by the nonselective evolution of postmating barriers, under allopatric or sympatric conditions; or, evolution of a postmating barrier as the primary event, followed by the selective evolution of premating barriers once the allopatric populations come into contact again. The latter alternative is not directly supported as there are no known incipient sympatric *Drosophila* species lacking premating isolation: the former alternative receives some support, although the requisite conditions for both the first and second steps of the sequence are unknown.

If there were information available on the way or ways in which premating barriers may evolve, whether sympatrically or allopatrically, whether selectively or nonselectively, or whether in various combinations of these conditions, then it might be easier to assess the speciation patterns which have given rise to the present-day sympatric species. For the present, the mechanisms involved must remain open to speculation. Probably, more than one mode has operated, as in the case of the allopatric species. Within any one closely related set of species, involving both sympatric and allopatric forms, there may well be examples of all the speciation modes found in Hawaiian flies, such has been the level of evolutionary opportunity and resulting divergence in this insular environment.

v GENETIC DIFFERENTIATION AND REPRODUCTIVE ISOLATION IN HAWAIIAN DROSOPHILA

Species are distinct systems, distinct because of genetic reproductive isolation. Species differentiation necessarily entails genetic differentiation, but only part (and perhaps only a small part) of this differentiation may be directly concerned with the establishment of the reproductive barrier. The reproductive isolation might be a consequence merely of the total spectrum of genetic differentiation, or it might be a discrete character controlled by relatively few genes which might arise independently of other kinds of genetic differentiation. At one possible extreme is the situation where major genetic differentiation is suggested to be a prerequisite for the appearance of a reproductive barrier. At the other extreme is the hypothetical situation where virtually the only inter-population genetic differences are those directly involved in providing reproductive isolation, whether these be behavioral or any other kind of difference. Both these viewpoints have been confidently advanced and supported.

Since reproductive isolation can evolve in several different ways in association with different speciation modes, there need be no uniform level of genetic differentiation associated with the speciation process. Carson has recently advocated the view that major reorganization of the gene pool has been directly involved in the speciation of most Hawaiian flies, because of the genetic revolution presumed to be associated with the founder event (Carson, unpublished). It might be supposed then that species arising by processes other than 'founder-type' ones should demonstrate lesser degrees of genetic differentiation. The genetic events responsible for the origin of such species are much less radical ones, and there is no reason to suppose that a complete genetic reconstitution need be involved.

However, the amount of genetic differentiation involved in species divergence must be largely determined by the sequence of the two kinds of event—

i.e. whether adaptive differentiation precedes or follows the evolution of reproductive isolation. The following relationships might be expected. Speciation resulting from founder events, where the reproductive difference is primary (Carson 1970, 1971), should be associated (at least initially) with substantially less genetic differentiation between related forms than speciation resulting from geographical division, in which the reproductive barriers only evolve after adaptive differentiation. Genetic differentiation has two aspects—the overall total of allelic differences, and the organization of these differences. Various speciation patterns undoubtedly involve these two aspects to differing degrees.

Prakash (1972) provides data in support of the proposed founder sequence of speciation, involving evolution of reproductive isolation in the presence of only minimal levels of genetic differentiation (Carson 1970, 1971). The Bogotà population of *D. pseudoobscura*, which is completely isolated from the mainland U.S. populations of the species, shows one way F_1 male sterility and loss of viability in the backcross males. The hybrid male sterility can be explained by differentiation of the Bogotà population at only four genetic loci, two on the X chromosome and two on the autosomes. This population appears to be of very recent origin, and the result of a founder event. Prakash concludes that the reproductive isolation has arisen as a direct consequence of founder effects, and without incorporation of different alleles at many loci. It is not a byproduct of extensive genetic divergence.

The amount of genetic differentiation between Hawaiian *Drosophila* species can be crudely assessed from the hybridization data. Many inter-island hybridizations result in viable progeny, which show no significant disruption of their developmental and presumably genetic systems. Some of these inter-island hybrids are actually fertile (Yang and Wheeler 1969). Hybrids between the sympatric species *D. silvestris* and *D. heteroneura* are also fertile. In Hawaiian *Drosophila*, the genetic divergence associated with speciation events has often been minimal.

An indication of the degree of compatibility between the genetic systems of different Hawaiian species of the picture-winged group is given by the comparatively high values of the 'index of oogenesis, I_o' (Kambysellis 1970). This index gives a measure of the growth and development of ovaries implanted from one species into another and appears to correlate strongly with the phyletic distance between host and donor species (Kambysellis 1968), and in a general way with the success of interspecific hybridizations.

A more refined comparison of the degrees of genetic differentiation between Hawaiian species could be made using genetic similarity coefficients calculated from isozyme frequency data. This approach has greater application since hybrids need not be produced. Comparison of this kind of data (Johnson, in preparation) with the various proposed modes of speciation in the group might demonstrate whether there is any particular relationship between them. It would be important to use only closely related and recently derived species for the comparisons, because genetic divergence is expected to increase with time since species divergence. Various expectations could be tested, and with sufficient data, some general patterns may emerge.

The level of genetic differentiation and the kind of reproductive barrier established may be more or less interdependent. Male hybrid sterility is very frequent between Hawaiian species; however, it seems to be present between species arising via various modes. It is possible that the observed hybrid sterility may have a different genetic basis in different situations. For example, in the founder-type situation, male sterility may result from a few specific genes (*cf.* Prakash 1972); in the case of the 'population splitting' situation, presumed to apply to the forms of the Maui-complex, hybrid sterility may result from a general incompatibility related to total genetic divergence. This is mere speculation at this stage, but the possibility should be kept in mind in interpreting instances of hybrid sterility, and reproductive barriers in general.

The isolation between many Hawaiian species is only partial, and for this reason their validity as species has been questioned (*cf.* Yang and Wheeler 1969). In some respects, the forms studied do not meet the 'species' criteria applied elsewhere. This is more evident amongst some of the allopatric forms which are apparently separated only by partial postmating barriers. Yang and Wheeler (1969) and Carson *et al.* (1970) conclude that complete reproductive isolation may not always accompany speciation. The incompleteness of some of the barriers between Hawaiian flies may merely be due to the recency of their divergence: in other cases, the partial genetic barriers are in effect complete by virtue of the presence of the geographic barriers. In any event, with time, most species would probably become completely isolated reproductively, as genetic differentiation within each proceeded and divergence increased. The great variety of degrees of reproductive isolation found between Hawaiian *Drosophila* species is a consequence of the unique biological situation in the Hawaiian Archipelago. As such, it offers unparalleled opportunities for the study of the evolution of reproductive barriers, and furthering of our understanding of speciation processes.

VI SUMMARY AND CONCLUSIONS

This paper attempts to assess the nature of the reproductive barriers found in association with the recognized specific differentiation in Hawaiian *Drosophila*, and to relate this information to an interpretation of insular speciation. The available data, predominantly from interspecific hybridization studies, indicate that Hawaiian *Drosophila* display great variation in the kind and degree of reproductive isolation between species, and that this variation is not necessarily related to their degree of phyletic relationship. The lack of any consistent pattern in the reproductive isolation of Hawaiian flies is a result of several factors: (1) the species differ in their ages, i.e. in the times since their initial differentiation, because of the unique interaction of biological, geological and geographical factors in the Hawaiian Islands; (2) there are several modes of species formation operative; and (3) the rates of specific divergence are highly variable, even within one particular speciation mode.

One of the most important conclusions is that not one, but several patterns of species formation have been involved in the evolution of the Hawaiian *Drosophila*. The explosive nature of the situation may be partly due to this fact—that insular conditions provide many favorable opportunities for the simultaneous differentiation of species via *multiple* modes, whereas the continental situation may be much more restrictive and may permit only *one* speciation mode in a given group over a given time period. Some of the speciation patterns which might be operative in the Hawaiian situation have been outlined, on the basis of the patterns of differentiation found between species showing various geographic relationships. By a comparison of the degrees of reproductive isolation between recent and relatively old species pairs, it has been possible to determine the nature of the primary reproductive barriers in each situation, and the sequence in the evolution of complete reproductive isolation.

The speciation of the allopatric forms is very definitely dependent on the existence of geographic barriers. Indeed, the importance of geographic isolation in the total evolution of this group cannot be overestimated—it is the one factor which has made the greatest contribution to the opportunity for species formation, both on a major and a minor scale. Between allopatric species, the primary reproductive barriers are postmating ones, usually involving sterility of hybrids (especially male sterility), and sometimes also partial or complete inviability of hybrids. These postmating barriers result from the genetic differentiation which usually accompanies allopatric separation of populations. Any premating barriers are necessarily secondary, and a byproduct of the total divergence between long-separated populations. The three different distribution patterns of allopatric species in Hawaiian *Drosophila* are probably associated with three different ways in which allopatric species might have been differentiated. These patterns are characterized by one or other of the following events: (*a*) founder event after an inter-island migration; (*b*) population splitting of one originally continuous population; (*c*) insular speciation unrelated to the founder event. The characteristics and the genetic differentiation associated with each of these proposed patterns have been discussed.

The evolution of sympatric species is harder to interpret than the evolution of allopatric species, but again this may proceed by more than one mode. Local geographic isolation of populations on the same island is a highly likely event because of the volcanic nature of the Hawaiian Islands, and this may often have been involved in the speciation of forms which are now sympatric. The primary reproductive barrier between some sympatric *Drosophila* species is a premating barrier probably determined by sexual behavioral differences. The impetus for evolution of such a barrier is an open question.

The fact that speciation can proceed by several different routes, and that it evidently has done so in the Hawaiian *Drosophila*, implies that there need be no general relationship between the amount of genetic differentiation and the speciation event. This controversy would be set to rest by recognition of this fact, and comparison of the contrasting modes of speciation within a group such as the Hawaiian *Drosophila*.

On the basis of their hybridization studies with the Hawaiian *Drosophila*,

Yang and Wheeler concluded that speciation is not necessarily accompanied by complete reproductive isolation. One alternative interpretation is that the forms recognized as species amongst Hawaiian *Drosophila* are not really valid species. It would be more reasonable to interpret the available data in terms of the total biological framework, and recognize that the range of degrees of reproductive isolation found is a result of the continuing nature of the speciation process, and the comparative recency of many of the critical phases of the process in a large proportion of the Hawaiian fauna. Reproductive isolation is certainly incomplete in many instances, but where the available mechanisms effectively preclude gene exchange under field conditions, the Hawaiian forms are no less valid species than continental species of *Drosophila* which evolved in the much more distant past. The picture-winged group of Hawaiian *Drosophila*, by virtue of their recency and phenomenal variation, provide an unparalleled opportunity for study of the processes of speciation under insular conditions, and continuation of this research should eventually yield some answers in the puzzle of evolution.

VII ACKNOWLEDGEMENTS

The author is sincerely grateful to colleagues associated with the Hawaiian *Drosophila* Project for much useful and stimulating discussion, and for permission to cite some of their unpublished data. Thanks are also extended to Professor S. Smith-White for his comments and criticisms on the manuscript draft, and to Mrs. E. Lockwood for typing the paper. The research was supported by N. S. F. Grant GB-29288 for Evolution of Hawaiian Drosophilidae Project.

REFERENCES

Carson, H. L. 1966. XIII Chromosomal races of *Drosophila crucigera* from the Islands of Oahu and Kauai, State of Hawaii. Univ. Texas Publ. **6615**: 405-412.

Carson, H. L. 1970. Chromosome tracers of the origin of species. Science **168**: 1414-1418.

Carson, H. L. 1971. Speciation and the founder principle. Stadler Symposium **3**: 51-70.

Carson, H. L., Clayton, F. E. and Stalker, H. D. 1967. Karyotypic stability and speciation in Hawaiian *Drosophila*. Proc. Nat. Acad. Sci. U.S. **57**: 1280-1285.

Carson, H. L., Hardy, D. E., Spieth, H. T. and Stone, W. S. 1970. The evolutionary biology of the Hawaiian Drosophilidae pp. 437-543. In: Essays in Evolution and Genetics in Honor of Theodosius Dobzhansky (M. K. Hecht and W. C. Steere, edd.), Appleton-Century-Crofts, New York.

Carson, H. L. and Sato, Joyce E. 1969. Microevolution within three species of Hawaiian *Drosophila*. Evolution **23**: 493-501.

Carson, H. L. and Stalker, H. D. 1968a. Polytene chromosome relationships in Hawaiian species of *Drosophila*. I. The *D. grimshawi* subgroup. Univ. Texas Publ. **6818**: 335-354.

Carson, H. L. and Stalker, H. D. 1968b. Polytene chromosome relationships in Hawaiian species of *Drosophila*. II. The *D. planitibia* subgroup. Univ. Texas Publ. **6818**: 355-365.

Carson, H. L. and Stalker, H. D. 1968c. Polytene chromosome relationships in Hawaiian *Drosophila*. III. The *D. adiastola* and *D. punalua* subgroups. Univ. Texas Publ. **6818**: 367-380.

Carson, H. L. and Stalker, H. D. 1969. Polytene chromosome relationships in Hawaiian species of *Drosophila*. IV. The *D. primaeva* subgroup. Univ. Texas Publ. **6918**: 85-93.

Clayton, F. E. 1968. XI. Metaphase configurations in species of the Hawaiian Drosophilidae. Univ. Texas Publ. **6818**: 263-278.

Clayton, F. E., Carson, H. L. and Sato, J. E. 1972. Polytene chromosome relationships in Hawaiian species of Drosophila VI Supplementary data on metaphases and gene sequences. Univ. Texas Publ. **7213**: 163-177.

Craddock, E. 1973. Interspecific hybridization studies in the *D. planitibia* subgroup of Hawaiian *Drosophila* (In ms.).

Dalrymple, G. B. 1971. Potassium—argon ages from the Pololu Volcanic Series, Kohala Volcano, Hawaii. Geol. Soc. Amer. Bull. **82**: 1997-2000.

Dobzhansky, Th. 1951. Genetics and the origin of species. 3rd Ed. Columbia University Press, New York.

Dobzhansky, Th. 1970. Genetics of the evolutionary process. Columbia University Press; New York and London.

Dobzhansky, Th. 1972. Species of Drosophila. Science **177**: 664-669.

Ehrman, L. (1962). Hybrid sterility as an isolating mechanism in the genus *Drosophila*. Quarterly Rev. of Biology **37**: 279-302.

Hardy, D. E. 1965. Insects of Hawaii. Vol. 12 Diptera Cyclorrapha II. Honolulu: Univ. of Hawaii Press, pp. 814.

Hardy, D. E. and Kaneshiro, K. 1972. New picture-winged *Drosophila* from Hawaii, Part III (Drosophilidae, Diptera). Univ. Texas Publ. **7213**: 155-161.

Heed, W. B. 1971. Host plant specificity and speciation in Hawaiian Drosophila. Taxon **20** (1): 115-121.

Kambysellis, M. P. 1968. Interspecific transplantation as a tool for indicating phylogenetic relationships. Proc. Nat. Acad. Sci. U.S.A. **59**: 1166-1172.

Kambysellis, M. P. 1970. Compatibility in insect tissue transplantations. I. Ovarian transplantations and hybrid formation between *Drosophila* species endemic to Hawaii. J. Exp. Zool., **175**: 169-180.

Kaneshiro, K. Y. 1969. V. A study of the relationships of Hawaiian *Drosophila* species based on external male genitalia. Univ. Texas. Publ. **6918**: 55-70.

Kaneshiro, K. Y., Carson, H. L., Clayton, F. E. and Heed, W. B. 1973. Niche separation in a pair of homosequential Drosophila species from the island of Hawaii. Amer. Nat. **107**: (in press).

Mayr, E. 1942. Systematics and the origin of species. Columbia University Press, New York.

Mayr, E. 1954. Change of genetic environment and evolution. In: Evolution as a Process. J. Huxley, Hardy and Ford, edd., pp. 157-180.

Mayr, E. 1963. Animal species and evolution. Belknap Press of Harvard Univ. Press. Cambridge, Mass.

Mayr, E. 1970. Populations, species and evolution. Harvard Univ. Press. Cambridge, Mass.

McDougall, I. 1964. Potassium-argon ages from lavas of the Hawaiian Islands. Bull. Geol. Soc. Amer. **75**: 107-128.

McDougall, I. 1969. Potassium-argon ages on lavas of Kohala Volcano, Hawaii. Bull. Geol. Soc. Amer. **80**: 2597-2600.

Patterson, J. T. 1942. Isolating mechanisms in the genus *Drosophila* Biol. Symp. **6**: 271-287.

Patterson, J. T. and Stone, W. S. 1952. Evolution in the genus Drosophila. New York, the Macmillan Company pp. 610.

Prakash, S. 1972. Origin of reproductive isolation in the absence of apparent genic differentiation in a geographic isolate of *Drosophila pseudoobscura*. Genetics, **72**: 143-155.

Rockwood, E. S., Kanapi, C. G., Wheeler, M. R. and Stone, W. S. 1971. X. Allozyme changes during the evolution of Hawaiian *Drosophila*. Univ. Texas Publ. **7103**: 193-212.

Spieth, H. T. 1966. Courtship behaviour of Hawaiian Drosophilidae Univ. Texas Publ. **6615**: 245-313.

Spieth, H. T. 1968. Evolutionary implications of sexual behavior in *Drosophila*. Evolutionary Biology, **2**: 157-193. New York, Appleton-Century-Crofts.

Stearns, H. T. 1966. Geology of the State of Hawaii. Palo Alto, Calif., Pacific Books.

Takada, H. 1966. Male genitalia of some Hawaiian Drosophilidae. Univ. Texas Publ. **6615**: 315-333.

Yang, H. Y. and Wheeler, M. 1969. XI. Studies on interspecific hybridization within the picture-winged group of endemic Hawaiian *Drosophila*. Univ. Texas Publ., **6918**: 133-170.

Effects of dispersal, habitat selection and competition on a speciation pattern of *Drosophila* endemic to Hawaii

R. H. RICHARDSON

Department of Zoology, University of Texas, Austin, Texas 78712

INTRODUCTION

Since the processes involved in speciation, particularly those of acquiring reproductive isolation, are elusive to experimental manipulation, we have examined natural populations seeking to observe actual cases of partial reproductive isolation which may eventually lead to speciation (e.g. the work reported by Dr. G. L. Bush in this volume) or to analyze recent speciation events *ex post facto* (e.g. the case in this paper). The discovery that two closely related species of Hawaiian *Drosophila, D. mimica* and *D. kambysellisi* occur in the same limited geographic area on the volcanic slopes of Mauna Loa has provided us with a genetically and ecologically analyzable example of recent speciation. Evidence has been accumulated by several colleagues and myself on diverse aspects of these species, including genetic data, habitat preferences, and dispersal patterns, which may be interpreted in light of the geologic and floral histories of the common habitat. These data suggest that the species may be an authentic example of sympatric speciation which can be experimentally verified, and constitutes a second chapter following the paper of Heed (1971).

It is not intended to present an argument favoring sympatric speciation over allopatric speciation as a general explanation for evolution among Hawaiian Drosophilids. Allopatric speciation has, doubtless, played a major role in the evolution of the Hawaiian Drosophilidae (Carson *et al.* 1970; Heed 1971). The objective, instead, is to examine a single case of two closely related species which occur in the same locality and, through analysis and comparison of all the genetic, behavioral, and ecological data at hand, demonstrate that these two species could, in fact, have evolved without having been geographically isolated.

Furthermore, our hypothesis implicates sympatric speciation (in addition to colonization) as a mechanism of 'species packing' (MacArthur 1971) during the evolution of a community. Although colonization clearly plays the key role of increasing the number of species on impoverished islands, the later stages of species packing may be through speciation. If so, community structure, optimal utilization of resources, and other ecological aspects of a biome may be recognized as products of the evolutionary processes of speciation and genetic adaptation.

THE THEORETICAL MODEL

Maynard Smith's (1966) theoretical model of potentially stable polymorphisms seems applicable to our case. His theoretical development shows that a stable polymorphism can exist between two alleles, even without sexual isolation, as long as each allele confers a selective advantage in one of two niches, and if the population sizes of the two morphs are regulated independently. That is, the niches and morphs must be sufficiently noncompetitive in their own niches that population size is not affected by the other morph. Also, for a simple model of one adaptive locus differentiating the morphs, the selective advantages must be sizeable. Smith further showed that incorporation of an additional genetically determined factor, such as habitat selection or assortative mating, allows the stable polymorphism to become the first stage of sympatric speciation. In its simplest state, only two loci need be involved, one for niche adaptation and one for behavior.

The prerequisites for Maynard Smith's model appear to be met in the case of the two closely related species, *D. mimica* and *D. kambysellisi*, both of which are endemic to the island of Hawaii. Expansion of the genetic model, which describes two loci each with two alleles, so as to encompass the considerable degree of genetic differentiation presumed between these two closely related species requires careful consideration of the factors involved in genetic differentiation. The disparity between the model and actual example can be reconciled by thinking of speciation as a process separate from adaptation. Actually, of course, the two processes are by no means mutually exclusive and, obviously, their effects are interrelated. For the purpose of our analysis, however, we should like to point out that they can occur independently. Two races may be formed sympatrically by accumulation of adaptive genetic changes at one or more loci according to the basic tenets of Maynard Smith's model. When behavioral variation, i.e., habitat selection, assortative mating, or both, represent part of the set of loci which are diverging, the process of speciation has begun.

It is a generally accepted tenet that the restriction of gene flow between segments of a population can result in their more efficient genetic differentiation, though the magnitude of the effect of such restriction is, as yet, unclear. If, however, gene flow were a significant factor retarding adaptation of the two races to their respective niches, then reproductive isolation would allow marked additional divergence and the resulting species would be much more highly differentiated than would be possible for the races. To properly weigh the effects that gene flow may have had on adaptation in the case of *mimica* and *kambysellisi* is beyond the purview of this paper. We will, instead, simply describe the differences we have observed to be associated with adaptation of the two species to their respective niches and allow the reader to draw his own conclusions as to the sequence of events which led to these differences.

Origin. The island of Hawaii is the most recently formed of the Hawaiian archipelago. This island is composed of five volcanoes, the oldest of which is less than one million years old (Stearns 1966). The oldest portion of the island is composed of Kohala Mountain, a greatly eroded volcano. The other volcanoes on the island are Mauna Kea, Hualalai, Mauna Loa and Kilauea. Haulalai, Mauna Loa and Kilauea have all erupted during recent history and Kilauea is presently active.

The area of sympatry and also the principal locality for both *D. mimica* and *D. kambysellisi* is the well-defined area of Kipuka Puaulu and Kipuka Ki on the southeast flank of Mauna Loa near its contact with Kilauea at between 1200 and 1400 meters above sea level (Figure 1). (A *kipuka* is an 'island' of older land completely surrounded by more recent lava flows.) Some kipukas, including those of this study, are heavily vegetated and contrast markedly with the relative nakedness of the recently deposited lava. A kipuka, surrounded by new lava, is thus isolated from undisturbed areas left untouched by the lava flow. Kipukas formed on old land surfaces may have soil layers and relatively deeply rooted vegetation, including large trees.

The development of isolated areas of vegetation does not necessarily imply the existence of a kipuka, however, as extensive areas of forest do not necessarily imply ancient undisturbed areas near the volcano. Smathers (1971, Ph.D. dissertation) has found adequate amounts of phosphate available for plant growth in fresh cinders as well as considerable amounts of sulfates and nitrates from dissolved volcanic gases in rainfall during eruptive activity. Furthermore, as seen in the areas of cinder and pumice deposition during the 1954 eruption of Kilauea Iki, so long as some large trees are not buried more than a meter or two, they may actually thrive on the additional nutrients from the deposits.

The surfaces of Kipuka Puaulu and Kipuka Ki are prehistoric ash deposits resting on Mauna Loa lavas. Kipuka Puaulu is surrounded on three sides by younger prehistoric Mauna Loa lava, and on the fourth side by lava which flowed from Kilauea, so that this kipuka is situated precisely at the trough between the slopes of these two volcanoes. Kipuka Ki, about a kilometer west of Kipuka Puaulu, is surrounded solely by late prehistoric lava from Mauna Loa (Peterson 1967). Mueller-Dombois and Lamoureaux (1967) have studied these kipukas extensively. From undated typescript of J. F. Rock, they report that the total soil depth in Kipuka Puaulu is approximately six meters. During their study, Mueller-Dombois and Lamoureaux found charcoal at a shallow depth of approximately 70 centimeters depth in both kipukas. C^{14}-dating of charcoal from Kipuka Ki yielded an age of 2,170 ±200 years.

These botanists made further observations of interest concerning the ages and origin of these kipukas. In their pits in Kipuka Puaulu, dug on level ground to a depth of two meters, they found lateral stability (equal depth on all sides of the pit) for the surface horizons above the depth of the charcoal, while there was lateral instability below the level of the charcoal. This pattern of strata could indicate the cinder deposits were being moved either by wind and/or water at stages of formation more than about 2,000 years ago, while they were stabilized more recently. Presumably the stability would result from extensive vegetation with adequate root systems to hold the cinder deposits in place. More data from other pits in the kipukas would be useful in evaluating their interpretation.

The charcoal layer can be helpful for dating the beginning of a suitable *Drosophila* habitat, which would have formed after an extensive fire. Thus 2,000

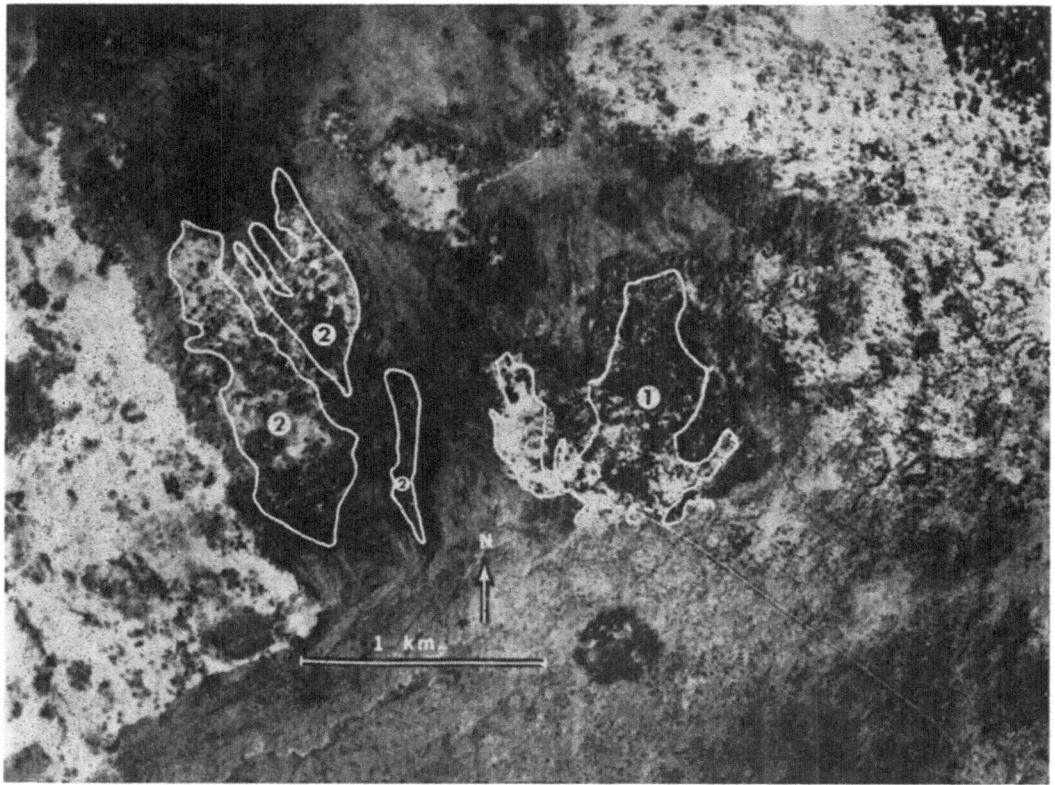

Figure 1. Aerial photograph of Kipuka Puaulu (1) and Kipuka Ki (2), with geological boundaries of the kipukas indicated by white bounds. Comparing this figure with Figure 8 gives a clear view of the position of lava flows of differing ages. Note the reinvasion of the lava flow, outside the geological boundary of the kipukas, by vegetation including trees. The lightest areas are primarily savannah, while darker more granular areas are forests. (Photo taken from number 0079 of the HAI series flown October 14, 1954, over the Hawaii Volcanoes National Park area. Geological boundaries supplied by Dr. Donald W. Peterson from original field notes and from published geological map (Peterson, 1967) after inspection of the aerial photo.)

years minus the length of time to grow the trees would estimate the beginnings of a *Drosophila* habitat. In addition, the present boundary of dense forest vegetation extends beyond the boundary of Kipuka Ki and Kipuka Puaulu onto the surrounding flows, (Figure 1), which suggests that the vegetation within the kipukas was likely destroyed by fires started from the freshly flowing lava and that reforestation has occurred on the edges of the flow to almost the same extent as inside the kipuka (Peterson, personal communication).

Peterson and Baker (personal communications) both are of the opinion that there was an older cinder blanket over this area which could have supported forest prior to the fire 2,000 years ago. Most of this blanket was covered by lava flows which formed Kipuka Puaulu and Kipuka Ki. Peterson feels that these flows were almost certainly later than the deposition of the stratum containing the charcoal formed about 2,000 years ago. Kipuka Puaulu was formed earlier than Kipuka Ki, since the surface lava around it is older than that around Kipuka Ki. In addition, Peterson indicated that the flows north east and west of the two kipukas are of such age that the vegetation supported by them also had to develop since the time the 2,000 year old charcoal was produced. Thus, these kipukas were initially separated from forested areas of any type by one to several kilometers, but forests have developed recently to the north east on some of these recent prehistoric lava flows. The forest in the kipukas is no more than 2,000 years old.

Characteristics. The woody vegetation in the kipukas consists mostly of native plants, although a few recent introductions may be found. Doty and Mueller-Dombois (1966) describe the principal habitats in Kipuka Puaulu either as 'open mixed *Acacia Koa-Sapindus* forest (or savannah) with *Metrosideros*', or a 'closed mixed *Acacia Koa-Sapindus* forest'. In the latter vegetation category the *Acacia Koa, Sapindus saponaria* and some *Metrosideros polymorpha* trees form the upper story of the forest, while there are numerous species of shade tolerant understory trees and arborescent shrubs (*Myrsine, Coprosma, Myoporum, Pipturus, Pisonia,* and others). In the forested area of Kipuka Ki there is less of an understory, and the *Acacia Koa-Sapindus* forest is more 'open'. The savannah regions in both kipukas are spotted with *Acacia Koa, Sophora chrysophylla* and *Sapindus saponaria* trees. The ground cover in the closed forest is principally leaf and fruit litter. However, in the open forest a number of native plants may be found growing close to the ground level, such as *Peperomia*,

Figure 2. Densities of *Drosophila* in the study area in Kipuka Puaulu. *D. mimica* (2 (a)), *D. kambysellisi* (2 (b)) and *D. imparisetae* (2 (c)) densities shown both as 3-D and contour plots, are given for the composite data over fifteen days of collections. (Edges of contour and 3-D plots are limits of sampling areas, not limits of the populations.) Three habitats are represented in the boundary overlays shown on the contour plots. The overlay on *mimica* is the boundary of the *Sapindus* canopy, within which will be found the fruit. The boundary on *kambysellisi* is that of *Pisonia* clones in the study area, within which will be found the rotting leaves. The boundary for *imparisetae* is that of saplings of *Sapindus*, which indicate an area of fruit fall with additional shading (thus the lower light intensity) from the understory. Note the nearly complete absence of overlap between *mimica* and *kambysellisi*, with *imparisetae* being located between.

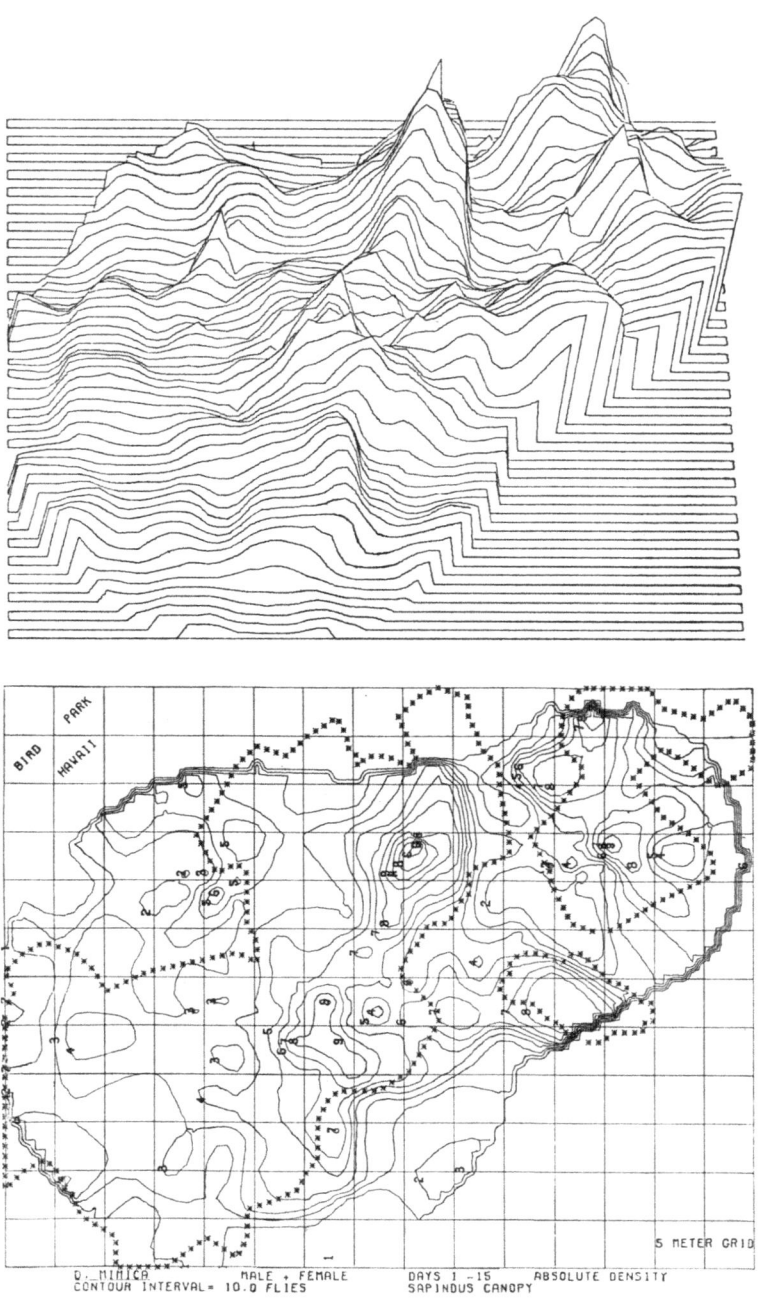

Figure 2(a)

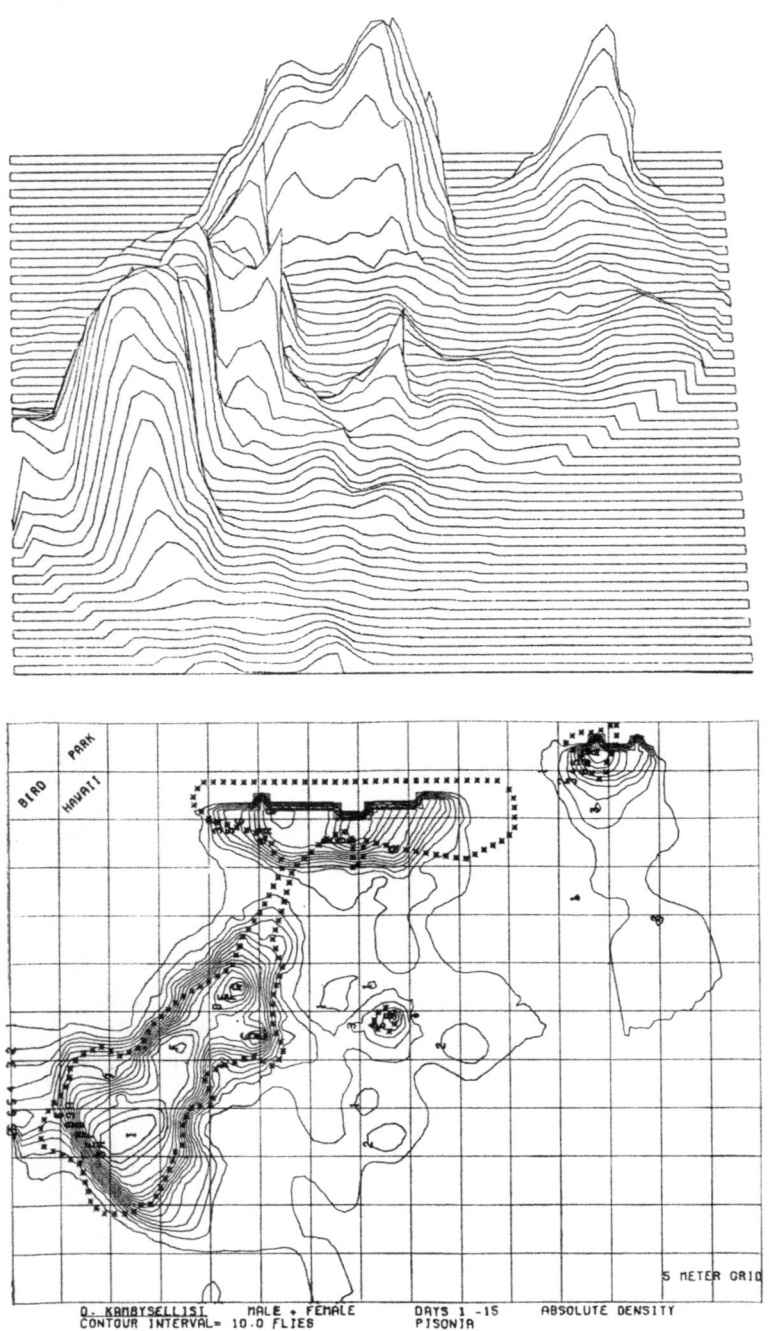

Figure 2(b)

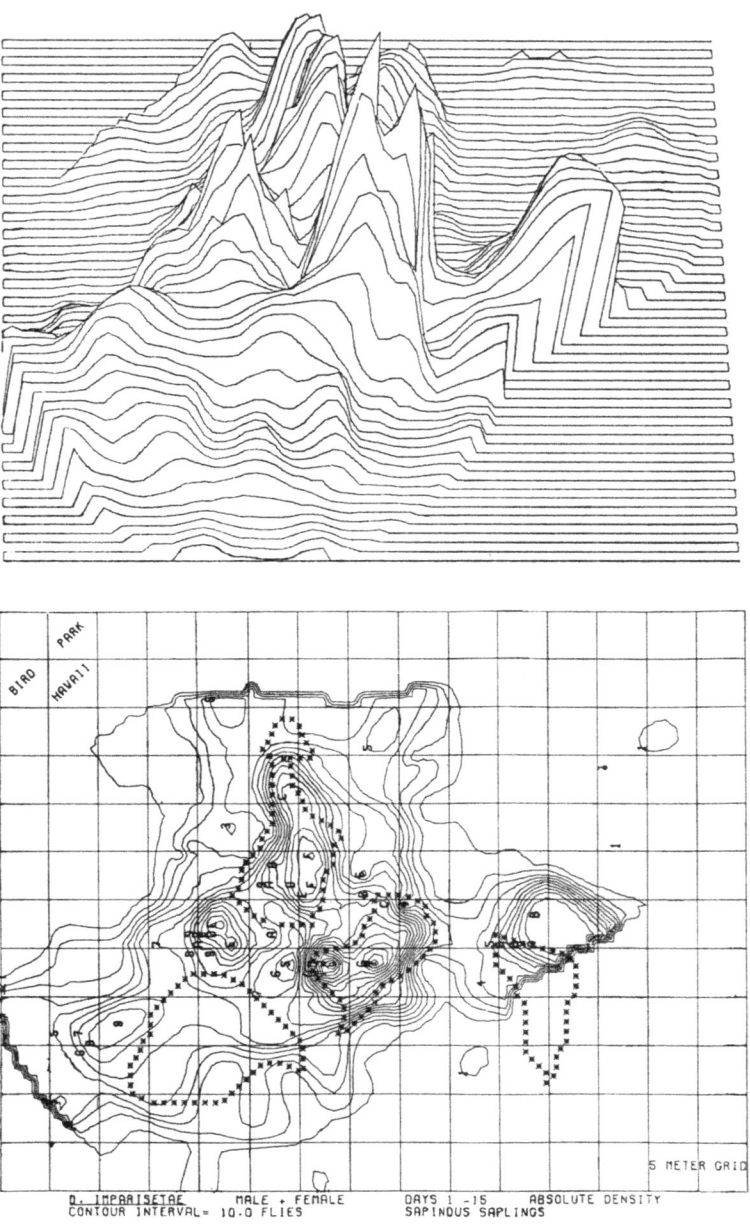

Figure 2(c)

Commelina, Microlepia, numerous grasses and wild strawberries (*Fragaria vesca* f. *alba*).

These two kipukas are relatively dry, situated on the saddle between Mauna Loa and Kilauea just over a ridge from the trade winds and therefore in a slight rain-shadow. A detailed analysis of the microclimates of these two kipukas by Smathers (1968) resulted in some interesting contrasts. Kipuka Ki received slightly more rainfall (33.4 inches *vs.* 31.4 inches), but with a greater evaporation rate during a study period from January to August, 1968. As a result of greater evapo-transpiration rate, Kipuka Ki appears as though it received less rainfall.

Although the rainfall for these kipukas averages between 50 and 75 inches per year, often there is reduced rainfall in the summer. During this period, however, clouds frequently pass through the kipukas. In a detailed study of micro-habitats of *Drosophila* made in June and July, 1969, the relative humidity at about 10 cm above the ground under both types of forest in Kipuka Puaulu was almost always above 95 per cent, but the relative humidity fluctuated daily, and fell to a low of about 65 per cent for brief periods at mid-day. *Drosophila mimica* was present in the open forest of *Sapindus* when the humidity was high, but the population disappeared from this area as the dry period persisted. As Smathers had observed in Kipuka Ki, the greater evapo-transpiration rate of the open forest allowed the soil and organic material at the surface to dry more rapidly than in a closed forest.

Since the kipukas are about 1,200 meters above sea level, they are cool. The temperature varies between 10°C and 24°C in the summer. Rare frosts have been observed in winter, but as a rule, 4°C is the minimum for a year and occurs most likely in February (Doty and Mueller-Dombois 1966).

Light intensity in summer varies widely, depending upon the canopy overhead. Reflected light from the ground cover on a clear day in the savannah of Kipuka Puaulu was measured at near 300 footcandles, while at the same time under the closed forest it was found to be as low as 2.5 footcandles. There are very sharp changes in light intensity associated particularly with the type of plants forming the understory. We recently found light intensity to be a particularly important *Drosophila* habitat parameter (Richardson and Spencer Johnston unpublished).

Heed (1968, and personal communication) found larval substrates for these species to be rotting leaves or fruits, with considerable species specificity as described for each species. The habitats of the plants supplying larval food substrate to these *Drosophila* species closely parallel the habitats of both adult and larval stages of the flies, as learned in recent studies (Richardson and Spencer Johnston, unpublished). *D. kambysellisi* females were rarely found away from rotting *Pisonia* leaves, although the males might be a meter or two away. The females of *imparisetae* were always in deep shade on the ground with *Sapindus* fruits. *D. mimica* was in more brightly illuminated areas of *Sapindus* fruit fall, such as the open *Sapindus* forest. This study (Figure 2) revealed strong habitat selection specific for the different species with overlap areas of no more than a few meters. The fact the *kambysellisi* was situated with *Pisonia,* which grows only in deep shade, while *mimica* was situated in open *Sapindus* forest resulted in no overlap of habitats of these species.

THE DROSOPHILA SPECIES

Three species are being considered as the primary components of the present *Drosophila* community found in a closed forest of *Sapindus* with *Pisonia*, although there are a few other species in low frequency in the areas studied. Two of the species in this community, *mimica* and *kambysellisi* are closely related and morphologically very similar members of the 'modified mouthparts' species group, and one, *imparisetae,* is in a miscellaneous group, but is similar in some respects to the 'white-tip scutellum' group (Throckmorton 1966). All are found only on the island of Hawaii.

D. mimica Hardy—This is the only one of the three species easily cultured in the laboratory, and consequently it is the best known. It has been collected from several localities on the island of Hawaii (Hardy 1965). Most of the specimens were collected in the Kilauea area, from Kipuka Ki and Kipuka Puaulu north east to the Upper Olaa Forest about 10 km away. Heed (1968) has reared *mimica* primarily from rotting fruits of *Sapindus saponaria*, but also from *Peperomia* leaves.

Clayton (1971) and Yoon (*et al.* 1972 and unpublished) have studied two of the close undescribed relatives of *mimica* and *kambysellisi* from the islands of Oahu and Kauai. Cytologically, the species on both Oahu and Kauai have heterochromatin added to the dot chromosome, giving a metaphase karyotype of six rods. The story probably is incomplete however, since recent collections and studies by Prof. M. R. Wheeler and Mr. Ken Kaneshiro (personal communications) indicate that each of these islands may have more than one 'mimica-like' species.

Clayton reported the metaphase chromosome complement of *mimica* to be six rods (Clayton 1966, 1968). There has been heterochromatin added to the usual 'dot' chromosome of the 'typical' primitive *Drosophila* karyotype of five rods and a dot chromosome, as well as one of the rod chromosomes, the X, has been lengthened by further addition of heterchromatin (Yoon, *et al.* 1972). Resch made crosses between *mimica* and the 'mimica-like' species on Oahu, but only sterile F_1 offspring were obtained, Yoon, *et al.* 1972).

D. mimica has not been observed mating in the field, although Spieth (1966) has studied the courtship behavior in the laboratory. The males are rather pugnacious, and show signs of territoriality, although they have not been observed to exhibit the lek behavior observed in numerous other Hawaiian *Drosophila.* Spieth describes the field behavior as one of recalcitrant females failing to respond to the courtship advances of the males as they both feed on the surface of the leaves. He found inseminations had occurred overnight in the laboratory studies, indicating that this species mates in the dark. Both sexes exhibit a photophilic response as the light level decreases in late afternoon (Figure 3). Mating is probably at night, high in the tree canopy, where the flies move at dusk.

Our studies of habitat selection have shown that *mimica* is located in areas

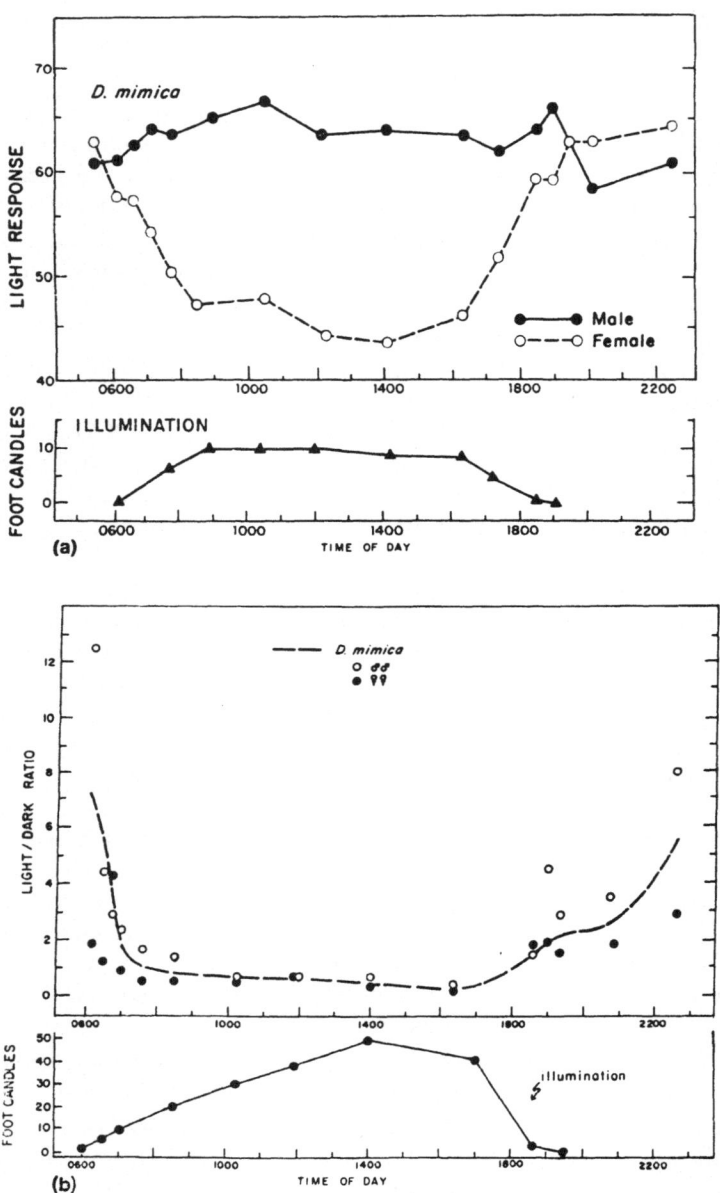

Figure 3. Photoresponse of *D. mimica.* Responses to low light intensity variations (a) and higher light intensities (b) are shown. The positioning in higher light intensity is indicated by increasing values on vertical axis in (d), while number of flies choosing total dark to number choosing light of various intensities is shown in (b). The males lack marked response to intensities up to ten footcandles, while the females show a marked response toward lower light intensity, situating themselves in less than five footcandles light.

of *Sapindus* fruit-fall where there is little or no understory. That is, this species is found either in an open *Sapindus* forest, or at the margins of a closed *Sapindus* forest. We have found that dispersal capabilities are adequate for the species to move among *Sapindus* separated by distances on the order of a kilometer (Richardson and Spencer Johnston, unpublished). Location of *Sapindus* fruit under isolated trees is efficient when the flies are downwind in low velocity air currents (Figure 4) as a result of the species being cued to fly upwind until in

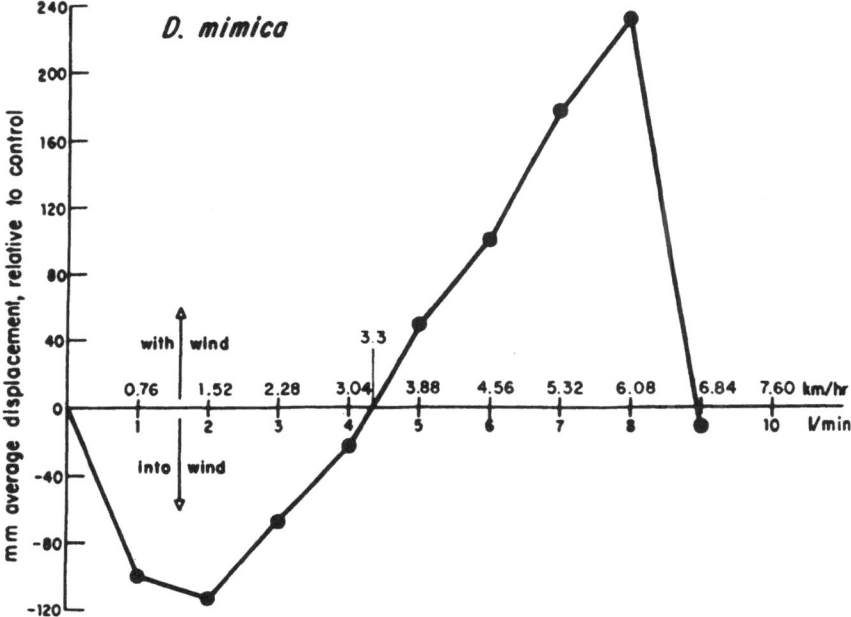

Figure 4. Response of *D. mimica* to air currents. Displacement of flies after three hours in an air flow, is shown relative to a control. Response is into the flow up to about 3.3 km/hr, then movement with the flow up to about 6.8 km/hr. At higher flow rates there is no response along path of air movement.

navigation range by olfaction to the habitat. At slightly greater velocities they would disperse downwind, while they show no directional response to still greater velocities. Presumably then they seek refuge from the wind.

   D. kambysellisi Hardy and Kaneshiro—This species is very similar externally to *mimica*. Both sexes are slightly smaller and darker colored than *mimica*. Kambysellis has found that internally the females of *kambysellisi* have about 7 or 8 ovarioles per ovary, compared to *mimica* which have about 12 per ovary (initially reported by Hardy and Kaneshiro 1969, and described in detail by Kambysellis and Heed 1971).

   D. kambysellisi and *mimica* are very closely related species (Hardy and Kaneshiro 1969). The polytene karyotypes of these species differ by only one fixed inversion although they differ more in metaphase karyotypes. In contrast to *mimica*, there is much less heterochromatin in metaphase chromosomes of *kambysellisi* (Yoon, *et al.* 1972, and unpublished data). Efforts to date here at Texas by Ms. Kathleen Resch (personal communication) have not produced any hybrids between *kambysellisi* and *mimica*-like species on the other islands, or with *mimica*.

This species has not been culturable in the laboratory until very recently. Dr. Linda Wheeler, with the help of Ms. Resch, has successfully cultured this species for four generations to date. Nevertheless, the species is much more demanding of specific culture conditions than is *mimica*. It has been reared only from rotting *Pisonia* leaves (Hardy and Kaneshiro 1969, and W. B. Heed, personal communication). Kambysellis and Heed (1971) have shown that the reduction in ovariole number further correlates with other leaf-breeding species, where apparently there has been reduced selection for ability to produce a rapid population expansion and, instead, there has been adaptation to more efficiently utilize a limited resource (K-selection of MacArthur and Wilson 1967).

The distribution of *kambysellisi* is extremely restricted. Prior to our specific search in 1972, the species had been found only in Kipuka Puaulu, but with the assistance of Mr. J. K. Baker, we found it in *Pisonia* growing in Kipuka Ki. Thus, so far as we know the species is located only in these two kipukas. Kaneshiro (personal communication) has specimens similar to either of *mimica* or *kambysellisi* from elsewhere on the island, but it is not known if they are relatives or represent new locations of these species. They were not collected near *Pisonia*.

We have not been able to demonstrate (unpublished data) that *kambysellisi* is cued to air movements when tested in the laboratory under the same conditions as was *mimica*, and the species is much less mobile between separated clones of *Pisonia* than is *mimica* for isolated *Sapindus* trees. In recent field studies, we never found *kambysellisi* movement out of a single stand of *Pisonia*. Furthermore, *kambysellisi* is very closely attracted to the *Pisonia* habitat. Even the males are not found far removed laterally from *Pisonia* leaf litter, although they are above the females by a meter or two. The species moves up into the canopy at night, presumably a phototropic response (Figure 5) to move toward the light (photophilic) in low intensity situations. The males also show a tendency to occupy more highly illuminated areas, such as above the ground on the top surface of leaves.

The mating behavior of *kambysellisi* is poorly known, except as suggested by the characteristics of the 'modified mouthparts' species group and *mimica* subgroup. From this morphological similarity we may assume that the mating behavior is similar to *mimica*, since the distinguishing morphological traits are largely a consequence of the male grasping the vaginal plates of the female during courtship (Spieth 1966). It exhibits lek behavior, and Spieth reports that field observations of *kambysellisi* (identified as *conjectura* in his 1966 paper, personal communication) revealed the males 'sitting on the edges of the leaves, 2 to 4 feet above the ground'. Further, he describes them as 'peering over the edges of these large leaves as if to watch the leaf litter below'.

Although there is no direct evidence of a system of chemical communication through pheromones in these species, Spieth (1966) describes male behavior of

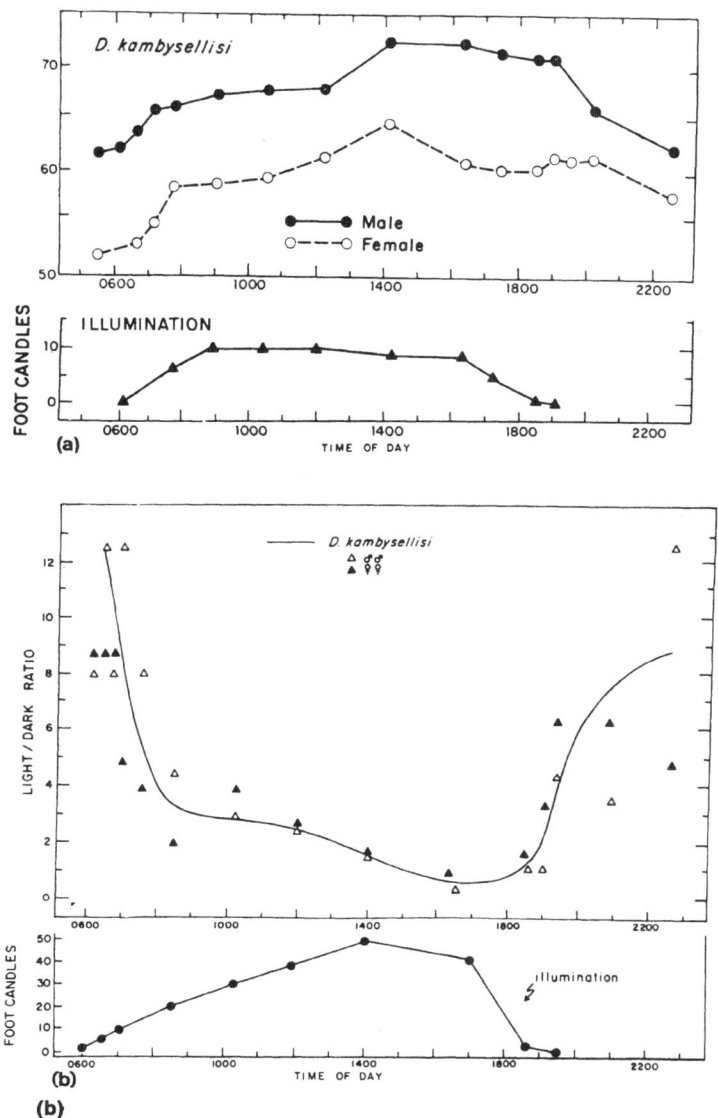

Figure 5. Photoresponse of *D. kambysellisi*. As in Figure 3, response in low light intensities (a) and higher intensities (b) is shown. This species tends to remain in lower intensities, nearly centred on ten footcandles, but with females consistently more photophobic than males. Higher light intensities strongly repel the flies of both sexes.

the 'picture wing' species group which suggests they are laying down pheromone trails by dragging the intra-anal organ across a small area. He and a student recently have found evidence of pheromones in lab tests (personal communication). Furthermore, Spieth suggests that the lek location may be chosen so as to intercept the females as they move away from the oviposition sites. As for *kambysellisi*, we found evidence for a species specific male attractant, (Figure 6)

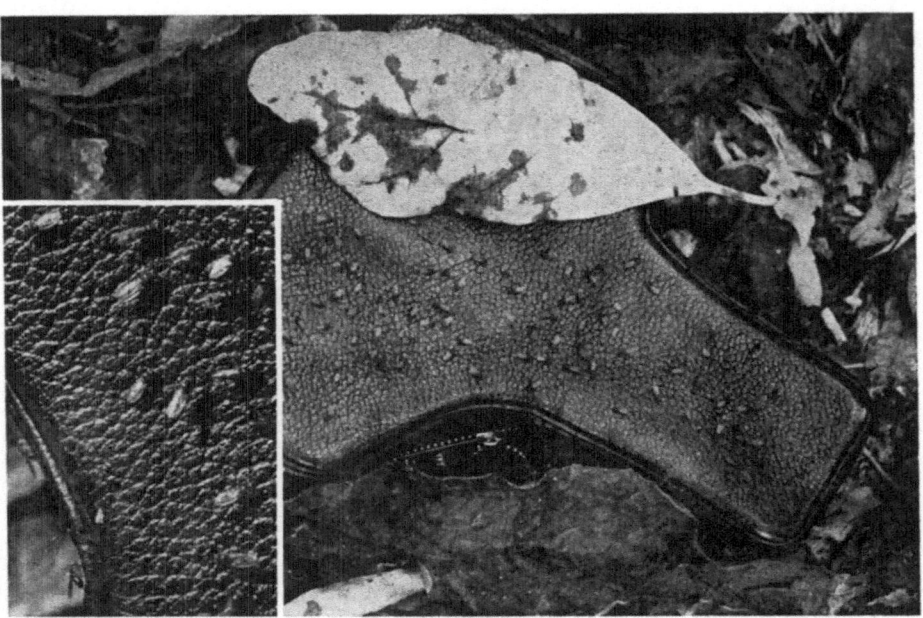

Figure 6. Photograph of a lightmeter case which was found to attract *kambysellisi* males (Johnston and Richardson, 1973). All flies on the case are *kambysellisi* males. Note that only three flies are on the *Pisonia* leaf placed across the case, and all are oriented toward the leaf edge toward the case! Insert shows lack of courtship activity and fairly regular spacing of individuals.

in the form of a leather lightmeter case (Johnston and Richardson 1973). We were able to collect several hundred *kambysellisi* males in a few minutes, without a single representative of another kind of fly. In addition, we were able to attract and collect a few *kambysellisi* males in an area several meters outside their usual habitats, where extensive local sweepings never had yielded a specimen of *kambysellisi*.

D. imparisetae Hardy—This species does not fit into the standard grouping of species in Hawaii, and consequently is placed in a miscellaneous category (Throckmorton 1966), although Hardy (1965) places it near the *haleakalae* complex of the 'white-tip scutellum' species group. Both of these systematists agree that this species is near the white-tip scutellum group, and Throckmorton (1966), using internal morphology as his criteria, places it near the phylogenetic branch point giving rise to the white-tip scutellum group. Thus it lies between this group and the 'modified mouthparts' species group.

Recently Ms. Kathleen Resch and Dr. Linda Wheeler succeeded in establishing, briefly, a few cultures from Kipuka Puaulu, allowing the cytological pattern to be examined. The species has six rods, confirming earlier unpublished findings of Clayton (personal communication from H. L. Carson). Yoon's results of cytological studies (personal communication) are consistent with Hardy's

(1965) and Throckmorton's (1966) placing this species between the 'modified mouthparts' group and the 'white-tip scutellum' group.

D. imparisetae has been collected in several places on Hawaii. Although genetic tests have not been performed (due to the inability to culture the species

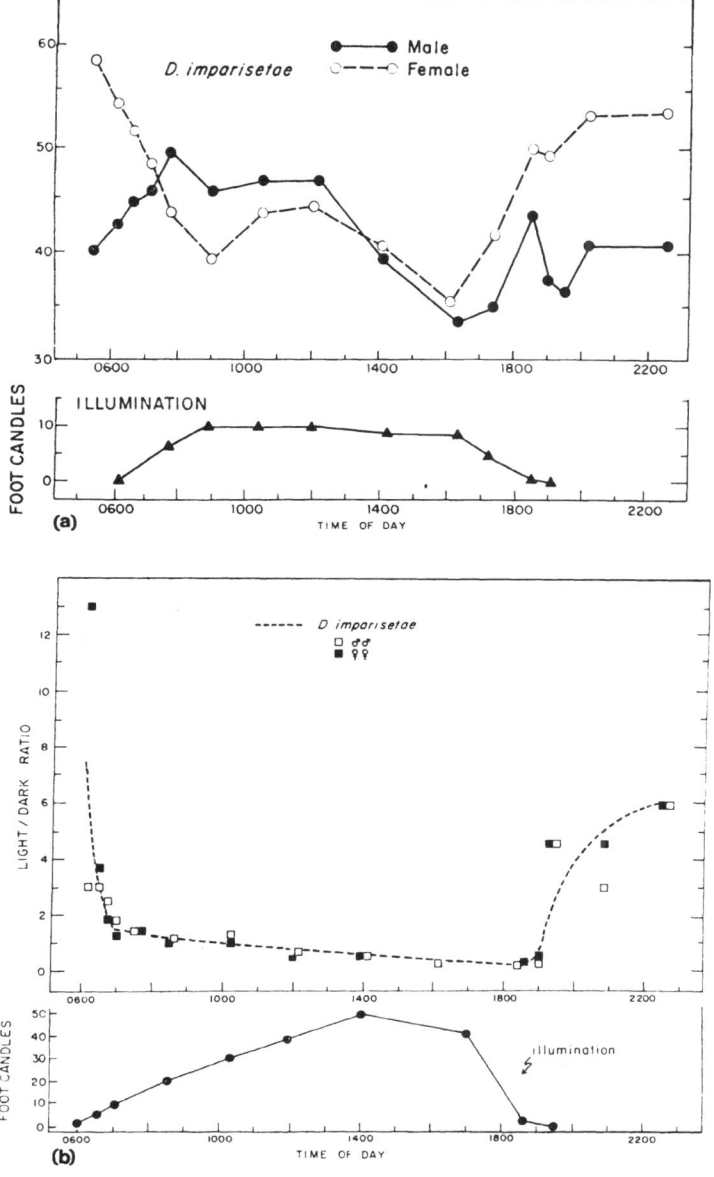

Figure 7. Photoresponse of *D. imparisetae*. As in Figure 3, responses in low light intensities (a) and higher intensities (b) are shown. Both males and females show a strong photophobic response to higher light intensities, but in low light intensities the males are less responsive than the females.

in the laboratory), specimens from several locations tentatively were classified as one species.

Spieth (1966) has described the mating behavior of *imparisetae,* both from laboratory and field observations. The species exhibits lek behavior, with the males tending to congregate in an area, but each male defending an individual territory. Often they were found around bracken fern and on *Pipturus* leaves, where the males were stationed on the underside of the leaves. Immature males would be found on the ground (Spieth, personal communication).

Heed (1968) reared *imparisetae* in large numbers from rotting *Sapindus* fruits collected in Kipuka Ki and from *Myrsine* fruit collected in the Olaa Forest area at Puu Hualalai; the habitat selection by *imparisetae* was distinct, and different from that of either *mimica* or *kambysellisi* (it consisted of heavily shaded areas of *Sapindus* fruit fall, adjacent to but not containing *Pisonia*). There was more spatial overlap between *imparisetae* and *mimica* than between *imparisetae* and *kambysellisi*. The positioning in heavily shaded areas agreed with the phototaxic response (Figure 7), where the species exhibited a strong aversion to even moderate illumination (photophobic). Nevertheless, the males were found in a more illuminated area than the females, agreeing with the field situation of a lek position on the underside of leaves above the ground.

THE ORIGIN OF A SPECIES AND THE DEVELOPMENT OF A NEW COMMUNITY

D. mimica is considered to be more primitive than *kambysellisi*. For example, it is ecologically a 'generalist', growing in an open forest of *Sapindus,* whereas *kambysellisi* is a 'specialist', growing under the lower canopy on *Pisonia* leaves. *D. mimica* is more widely distributed on Hawaii than is *kambysellisi*. It crosses with a *mimica*-like species on an older island, while *kambysellisi* does not. It occupies a habitat which could become available earlier in a plant succession found on a younger kipuka than the habitat of *kambysellisi,* and has a greater and more efficient dispersive capability than does *kambysellisi,* so that *mimica* could more effectively colonize new habitats as they emerged.

Even though *mimica* is more primitive in some aspects than is *kambysellisi,* it has a more 'derived' metaphase karyotype, having more heterochromatin than does *kambysellisi*. Either this karyotype (1) developed after *mimica* and *kambysellisi* speciated, or (2) there is an unknown common ancestor which is (was) *mimica*-like except for a primitive karyotype, or (3) there has been a loss of heterochromatin during the speciation leading to *kambysellisi,* or (4) *kambysellisi* is truly primitive and ancestral to *mimica*. (The first two possibilities are not mutually exclusive in the case of temporal speciation in the stem line.) The existence of *mimica*-like species on older islands also with six rod chromosomes suggests either common ancestral relationships, i.e., that premise (3) is the way the *kambysellisi* karyotype has been derived, or a number of other possibilities. Wheeler (personal communication), as well as Kaneshiro (personal communication), are of the opinion that at least one *mimica*-like and one *kambysellisi*-like species occur on each island, but the cytological examination has not been completed. (Kaneshiro has suggested the existence of a third part of the system, involving the *infuscata*-like species which are logically part of this general evolutionary complex.)

There are two patterns of evolution which would result in a series of *mimica*-like and *kambysellisi*-like species distributed down the chain of islands. The first would be divergence at the older end of the chain into a pair of lines, the *mimica*-like line and the *kambysellisi*-like line, followed by parallel colonizations and allopatric speciations leading to *mimica* and *kambysellisi* at the youngest end of the archipelago on Hawaii. Such a pattern was found by Carson and his colleagues in the 'picture wing' group where there has been parallel colonization of several of the subgroups (Carson, *et al.* 1970).

A second pattern would have a single ancestral line responsible for all inter-island colonizations with each colonization an event of allopatric speciation followed by small radiations on each island. This hypothesis deserves consideration since the divergence between the *mimica* and *kambysellisi* lines does not appear to increase from the older to the younger end of the island chain as would be predicted by the first hypothesis. In contrast to the variety of species involved in Carson's study of the 'picture wing' group, the *mimica* and *kambysellisi* species appear to be quite closely related and have, consequently, all been placed in a single *mimica* subgroup within the 'modified mouthparts' group (Yoon, *et al.* 1972). All are morphologically similar over the entire range from Kauai to Hawaii and there is considerable similarity of metaphase and polytene chromosome patterns. Furthermore, gene sequences in *mimica* and *kambysellisi* differ by only one inversion (Yoon, *et al.* 1972) and their electrophoretic patterns of proteins are very similar (Rockwood 1969), indicating that they have not accumulated the kinds of differences which would be predicted by the first pattern for two independently introduced species at the end of the phylogenetic line.

Although both models of the evolutionary pattern account for the same number of speciation events, the second model requires fewer inter-island colonizations, attributing the existence of more than one species of a single island to intra-island speciation events. Inter-island colonizations, obviously, fostered allopatric speciations, but at least some of the intra-island speciation could have been sympatric.

Application of either model to the actual evolutionary pattern of species of the *mimica* subgroup will have to remain conjectural until additional genetic, behavioral, and ecological comparisons can be made among species on individual islands and among analogous species in different radiations. For the moment, we will concentrate on events only at the phylogenetic terminal concerning *mimica* and *kambysellisi*.

As discussed earlier, *mimica* and *kambysellisi* are found in a forest habitat. The areas east, north and west of Kipuka Ki and Kipuka Puaulu are unsuitable *Drosophila* habitat, being sparsely covered by small *Metrocideros* trees and *Vaccinium reticulatum* ('ohelo'), *Dodonaea viscosa* ('aalii'), *Styphelia tameia-meiae* ('pukeawe'), and other areas are mostly prehistoric lava fields. Since much time has elapsed without any development yet of a closed forest cover, we can

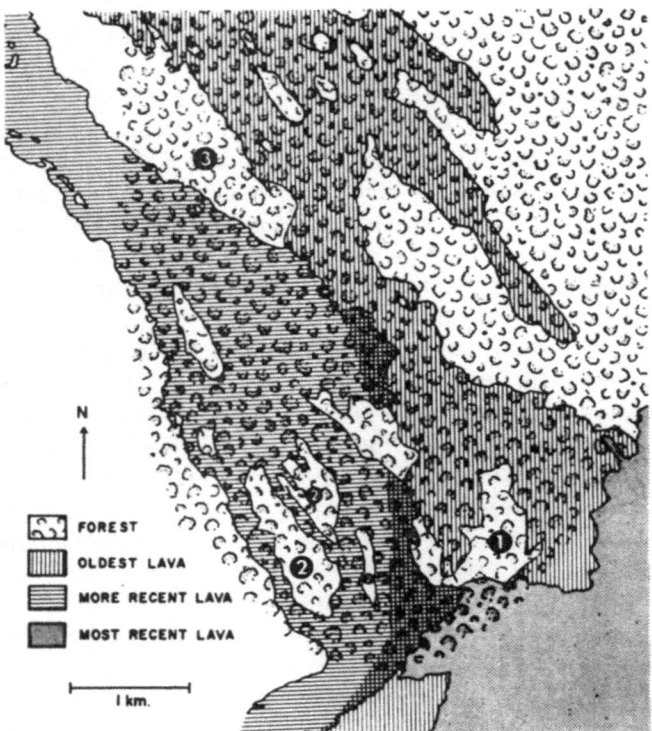

Figure 8. Drawing taken from geologic map (Peterson, 1967) showing hypothetical range of original forest area prior to surface lava flows. Kipuka Puaulu (1) is surrounded primarily by older lava, while Kipuka Ki (2) is surrounded by lava of more recent eruptions of Mauna Loa volcano. Most recent lava is from Kilauea volcano. Kipuka Kekake (3) is shown in the upper left of the figure.

assume that plant succession is progressing so slowly a *Drosophila* habitat has been absent for several hundred to a few thousand years.

D. *mimica* would readily colonize and persist in pre-climax succession of new kipukas consisting of an open forest of *Sapindus*. Its dispersal abilities would effectively guide it across sizeable lava fields, particularly since isolated *Sapindus* trees occur on relatively recent lava flows, where they have reinvaded the area outside the kipuka.

Both Dr. D. W. Peterson and Mr. J. K. Baker (personal communications) believe the areas within Kipuka Puaulu, Kipuka Ki, and possibly Kipuka Kekake may be remnants of a once larger forest (shown hypothetically in Figure 8). Indeed, it may have connected with what is now the Olaa Forest a few kilometers to the north-east where *mimica* has been collected.

Whether the kipuka tree population represents survival or reinvasion, the *mimica* habitat would occur earlier either in a plant succession reestablishing the kipuka vegetation or would more readily survive a fire than would the *kambysellisi* habitat. Peterson (personal communication) has suggested an alternative history of the forest in the kipukas. The rapid vegetative invasion of lava surrounding Kipuka Ki and the similarity of forests on recent and older lava suggest that Kipuka Ki is recovering from more recent vegetative devastation. Examination of Figure 1 shows clearly the reinvasion of surrounding lava flows by vegetation in both Kipuka Puaulu and Kipuka Ki. Baker (personal communication) favors the position that the vegetation on the two kipukas represents remnants of an older unconnected forest covering the general area of kipukas as well as surrounding recent prehistoric lava flows. Whichever point of view one takes, however, one reaches the conclusion that the present *mimica* habitat in the kipukas was established before that of *kambysellisi*. If *mimica* was present, it would have had a better chance of surviving volcanic disturbances than would *kambysellisi* because its habitat is more resistant to such factors as fire or partial burial by cinders. If destroyed, *mimica's* habitat would be reestablished quicker. The *Drosophila* would, naturally, be destroyed along with the habitat and would then have to recolonize the kipukas, in which case *mimica* has the best chance because its habitat is reestablished sooner and because it has more efficient dispersal abilities. It is highly likely that *mimica* would have become established in the kipukas before the niche of *kambysellisi* even developed.

Pisonia is readily spread by the fruits sticking to feathers of birds (Hillebrand 1965). Whenever introduced after the forest began to develop, it would grow in isolated clones as part of the understory. A mutation in *mimica* allowing it to use *Pisonia* as an egg deposition site would open the way for *mimica* to exploit this resource. The larval selection would be very different for such a race on *Pisonia* leaves compared with that on *Sapindus* fruit, and a stable polymorphism could result according to Smith's model. Flies carrying a mutation causing *Pisonia* to be attractive would be pulled well outside of the original habitat of *mimica*, quite analogous to a mutation affecting the origin of host-races in other insects. Additional adaptive changes brought about by differences not only in the food substrate, but also in other niche differences, certainly would occur.

Dr. Guy L. Bush discusses in this volume the origin of a new host race in *Rhagolitis pomonella* in only a few generations. The Tephritidae mate only on their specific host plant at the time of oviposition. A number of investigations on these true fruit flies (Christenson and Foote, 1960; Bush, 1965, 1966, 1969; and Prokopy and Bush, 1972) indicate rapid formation of host races without geographic isolation. Heed (1971) and Prof. D. E. Hardy (personal communication) point out that, within the Hawaiian Drosophilidae, *Exalloscaptomyza* has exploited a niche of flowers, particularly the indigenous species of *Ipomoea* spp. (morning glory) found at higher altitudes. Such a change most likely occurred rapidly, and was with specific associations between the species of fly utilizing a particular species of *Ipomoea*. These observations closely parallel those made by Bush and his colleagues.

There seems to be a close parallel between evolutionary patterns in these other insects and a logical interpretation of the events in the *Drosophila*. Maynard Smith's model for speciation is met if either assortative mating or habitat selection is present. In this case with *Drosophila*, both behavioral responses would be found, since mating is near the oviposition sites.

The lek mating behavior requires that the males must communicate over a distance with the females of their species. It seems that this is accomplished primarily through a chemical communication system of pheromones. Spieth's studies (1966) of lek behavior in *kambysellisi* revealed the males 'sitting on the edges of the leaves, 2 to 4 feet above the ground'. The females are attracted to the rotting leaves on the ground. Thus, it seems two processes are involved, differently expressed in each sex, and possibly under separate genetic control. The females respond to substrate attractiveness, and the males to female attractiveness. Since pheromonal communication can be effective over distances much greater than those separating the *Pisonia* and *Sapindus* habitats (Wright 1958), the assortative mating would not be as strong if both races used the same pheromonal signals. Thus, it seems that two changes are needed to fully accomplish positive assortative mating, first an attraction to the *Pisonia* (habitat selection) and, second, a change in pheromone signals between the races allowing sexual selection.

The simplest expectation, of course, were it not for our peculiar lightmeter case, would be where one locus effected a change in substrate attractiveness with common expression in both sexes. However, since we observed that only males were attracted, the suggestion is very strong that the leather in the case contained either, by a remote coincidence, the female pheromone of *kambysellisi*, or, more likely, a very similar chemical which mimics the pheromone or possibly serves as a male aggregation pheromone. It seems, therefore, that serious consideration must be given to the more complex genetic model, which involves at least a two locus change to achieve strong positive assortative mating.

If males are attracted to females via a female pheromone, rather than to substrates, then, the efficiency of the system would be increased by a high genetic correlation between a locus (or loci) causing *Pisonia* to be attractive to females and the pheromone allele causing these particular females to be attractive to the males carrying (though not expressing) the alleles for *Pisonia* attractiveness. The most obvious mechanism for maintaining a particular combination of non-alleles is to have them on the same chromosome with restricted recombination between the loci. Unless recombination is absolutely prevented, there must be selection against individuals carrying the recombination gametes (Kojima and Kelleher 1961, and many others since), but as linkage increases, the necessary selection intensity is reduced. However, linkage can become nearly complete through the incorporation of both loci into an inversion.

If an inversion polymorphism differentiated the population into races occupying different niches *and* caused positive assortative mating to develop, heterozygotes would rarely form and isolation between the races would be greatly increased. It would increase more rapidly than expectations under Maynard Smith's model, thereby accelerating the speciation process. Coincidence or not, an inversion differentiates *mimica* and *kambysellisi* gene sequences.

In addition, the present genetic diversity between *mimica* and *kambysellisi* may reflect the action of forces imposed by competition with a third species which has niche requirements intermediate between those of *mimica* and *kambysellisi*. In Kipuka Puaulu and Kipuka Ki the habitat intermediate between *Pisonia* leaves and *Sapindus* fruits is occupied by *Drosophila imparisetae*. At the time a new *mimica* race was forming on *Pisonia* leaves, hybrids between the new race and the *Sapindus* race of *mimica* would probably have been competing with *imparisetae* in the intermediate habitat. It is unlikely that novel genotypes such as the hybrids would have been able to outcompete a species already adapted to the habitat. To the extent that the niches of the hybrids and *imparisetae* overlapped there would have been selection against the hybrids. Since the physical location of *imparisetae*'s habitat in between that of the proposed races of *mimica*, competitive selection against racial hybrids would have reinforced spatial isolation of the races.

In summary, the scheme proposed for the origin of *kambysellisi* is as follows:

(1) *D. mimica* colonized or survived in a *Sapindus* habitat in the kipuka(s).

(2) Plant succession proceeded, finally developing a canopy forest including *Pisonia* in the understory.

(3) A new race of *mimica* developed to exploit the new resource provided by rotting *Pisonia* leaves.

(4) The new resource became attractive to males as well as females and positive assortative mating developed within races.

(5) Isolation, from habitat selection was reinforced because inter-racial hybrids were being eliminated in the intermediate habitat by competition with *D. imparisetae*.

(6) Isolation became complete (probably aided by the addition of a pheromonal recognition modification differentiating the two races, allowing sexual selection to operate) and a new species, *D. kambysellisi*, was formed.

The event of speciation clearly does not imply that the *mimica* and *kambysellisi* genetic repertoire seen today is the same as that which existed then. Certain kinds of adaptive genetic modifications might be possible only with complete isolation. For example, many loci are involved in a modification, but each with relatively small selective effects, gene flow could swamp the effects of selection. Furthermore, if the chance events which led to gene combinations resulting in isolation occurred quickly relative to the many changes ultimately associated with evolution of a new niche, then genetic divergence could be largely after speciation.

The data presented by Dr. Bush strongly suggest that isolation may occur very early in a diverging sequence of populations. Speciation, *per se*, need not necessarily involve a massive genetic modification, although an examination of genetic differences between closely related species at a single point in time generally reflects genetic differences involving more than a few loci.

We can state a few corollaries of this separation of genetic changes directly causing isolation and those related otherwise to adaptation. If speciation occurs early in a divergence, the ultimate genetic differences may reflect to a greater extent the differences in niches between the two species than if isolation were delayed. In the conceptual framework of Sewall Wright's adaptive surfaces, gene flow would tend to increase the chances of the new species occupying a peak nearer that of ancestral species, than if isolation occurred easily and divergence were unrestricted by gene flow (e.g., as in allopatry).

Another corollary is related to the types of differences between closely related species. If the differences are largely a result of adaptation, rather than chance, then they reflect in a general way the differences in niches of the derived and ancestral species. In the case of *mimica* and *kambysellisi,* we would conclude that they are very similar morphologically because external selective factors are similar. For example, they may share similar selective forces from predators so that camouflage is similar. (In fact, the slightly darker colored *kambysellisi* are usually sitting on darker substrate, rotting *Pisonia* leaves in a deep shade, than are *mimica,* which are sitting on rotting fruit and lighter colored leaves in the open shade.) On the other hand, the *mimica* reproductive system is greatly different from that of *kambysellisi*; there are almost twice as many ovarioles per ovary in *mimica.* Considering the reproductive differences associated with oviposition substrates for leaf-breeding species and other species found by Kambysellis and Heed (1971) the divergence in this way would be understandable which would be intrapopulation selection for reduced r, rate and population increase. Furthermore, with a similar genetic heritage, species evolving separately into similar niches would be expected to show at least parallel, if not convergent morphological modifications. In addition, once a species becomes adapted to a niche (i.e., reaches an adaptive peak) it should have long term genetic stability, paralleling the niche stability to a considerable degree, so long as major genetic perturbations are absent. (It should not be surprising to find 'relic' species in niches which were stable over long time spans.)

A third corollary relates to the evolution of a community. If a species can evolve *in situ* where there is an unexploited resource available then species packing may be accomplished by speciation as well as colonization. If such a phenomenon exists, then clusters of phylogenetically closely related species should be found in close geographic proximity. This process of species packing by speciation would result in a pattern of different resources being utilized by closely related species, whereas species packing by colonization would lead to the comparable resource being utilized by a great diversity of organisms. It is not surprising that the role of colonization would at first appear the most important, because the identification of species in a community would be easily accomplished. In the case of clusters of sibling species, however, their identification would be possible only where the forms in a community were well known, particularly having been studied by the geneticist and ecologist together. Only in a few instances, such as in the study of some of the Drosophilidae, are we in a position to observe at the present time the suggestions of species packing by speciation. The case of *mimica* and *kambysellisi* is one of those instances.

ACKNOWLEDGEMENTS

Many people have contributed to the ideas in this paper. Members of the Hawaiian Drosophilidae Project are conspicuous contributors, particularly M. R. Wheeler, Linda Wheeler, D. E. Hardy, H. L. Carson, W. B. Heed, H. T. Spieth, Michael Kambysellis, Kenneth Kaneshiro, Jong Sik Yoon, Stephen Montgomery, Kathleen Resch and John P. Murphy. They have freely supplied data, advice and criticisms. The recent collaboration with Dr. J. Spencer Johnston has been most fruitful, and material from our unpublished manuscripts has been freely used. Dr. Linda Wheeler has been a most helpful scientific editor, critic and illustrator. The extensive computer graphics in the analysis of the habitat selection and dispersal studies, of which only a selected portion is presented in this paper, was ably and elegantly performed by Villy M. Sorenson. Visualization of the pattern complexities would have been much more difficult, if not impossible, without his assistance. A few discussions with my colleagues Guy Bush and Daniel Udovik were most helpful.

The sections on geologic and ecologic history of the Kipuka Ki-Kipuka Puaulu area were possible only through the generous and enthusiastic assistance of Dr. Donald W. Peterson, Director of the U.S.G.S. Hawaii Volcanoes Observatory, and Mr. James K. Baker, Research Biologist at the Mauna Loa Field Station, Hawaii Volcanoes National Park. I have used freely from their correspondence, and considered many helpful suggestions they made on drafts of the manuscript sent to them for review. Of course I assume the responsibility for any mistakes which remain.

This research has been supported primarily by U.S. Atomic Energy Commission Contract AT-(40-1)-4023 and N.I.H. Career Development Award 5 K04-GM 47350, and secondarily by N.I.H. Grant GM 19616 and N.S.F. Grant 22770. Indirect support was given by an N.S.F. Grant to D. E. Hardy and N.I.H. Grant GM 11609 to M. R. Wheeler.

REFERENCES

Bush, G. L. 1965. The genus *Zonosemata* with notes on the cytology of two species. Psyche **72:** 307-323.

Bush, G. L. 1966. Taxonomy, cytology and evolution of the genus *Rhagoletis* in North America (Diptera, Tephritidae). Bull. Mus. Comp. Zool. **134:** 431-562.

Bush, G. L. 1969. Sympatric host race formation and speciation in frugivorous flies of the genus *Rhagoletis* (Diptera, Tephritidae). Evolution **23:** 237-251.

Carson, Hampton L., Hardy, D. Elmo, Spieth, Herman T. and Stone, Wilson S. 1970. The evolutionary biology of the Hawaiian Drosophilidae. In Essays in Evolution and Genetics Max K. Hecht and William C. Steere, edd. Appleton-Century-Crofts, New York.

Christenson, L. D. and Foote, R. H. 1960. Biology of fruit flies. Ann. Rev. Ent. **5:** 171-192.

Clayton, Frances E. 1966. Preliminary report on the karyotype of the endemic Hawaiian Drosophilidae. In Studies in Genetics III, M. R. Wheeler, ed. Univ. Tex. Pub. **6615:** 397-404.

Clayton, Frances E. 1968. Metaphase configurations in species of the Hawaiian Drosophilidae. In Studies in Genetics IV, M. R. Wheeler, ed. Univ. Tex. Pub. **6818:** 263-278.

Clayton, Frances E. 1971. Additional karyotypes of Hawaiian Drosophilidae. In Studies in Genetics VI, M. R. Wheeler, ed. Univ. Tex. Pub. **7103:** 171-181.

Doty, M. S. and Mueller-Dombois, D. 1966. Atlas for Bioecology Studies in Hawaii Volcanoes National Park. Hawaii Botanical Science Paper No. 2, University of Hawaii, Honolulu. 507 pp.

Hardy, D. E. 1965. Insects of Hawaii. Vol. 12, Diptera: Cyclorrhapha II, Series Shizophora, Section Acalypterae I, Family Drosophilidae. University of Hawaii Press, Honolulu. 814 pp.

Hardy, D. E. and Kaneshiro, K. Y. 1969. Description of new Hawaiian *Drosophila*. In Studies in Genetics V, M. R. Wheeler, ed. Univ. Tex. Pub. **6918:** 39-54.

Heed, W. B. 1968. Ecology of the Hawaiian Drosophilidae. In Studies in Genetics IV, M. R. Wheeler, ed. Univ. Tex. Pub. **6818:** 387-419.

Heed, W. B. 1971. Host plant specificity and speciation in Hawaiian *Drosophila*. Taxon **20:** 115-121.

Hillebrand, W. F. 1965. Flora of the Hawaiian Islands, Hafner Publishing Co., New York. 673 pp.

Johnston, J. S. and Richardson, R. H. 1973. A unique sex-specific, species-specific attractant for a Hawaiian *Drosophila*. Dros. Infor. Serv. **49**: 46.

Kambysellis, M. P. and Heed, W. B. 1971. Studies of oogenesis in natural populations of Drosophilidae: I. Relation of ovarian development and ecological habitats of the Hawaiian species. Amer. Nat. **105**: 31-49.

Kojima, K. and Kelleher, T. M. 1961. Changes of mean fitness in random mating populations when epistasis and linkage are present. Genetics **46**: 527-540.

MacArthur, R. H. 1971. Geographical Ecology. Harper and Row, New York. 269 pp.

MacArthur, R. H. and Wilson, E. O. 1967. The Theory of Island Biogeography. Princeton University Press, Princeton, New Jersey. 203 pp.

Maynard Smith, J. 1966. Sympatric speciation. Amer. Nat. **100**: 637-650.

Mueller-Dombois, D. and Lamoureaux, C. H. 1967. Soil vegetation relationships in Hawaiian kipukas. Pacific Sci. **21**: 286-299.

Peterson, D. W. 1967. Geologic map of the Kilauea Crater quadrangle Hawaii. U.S. Geological Survey. Map GQ-667.

Prokopy, R. J. and Bush, G. L. 1972. Mating behavior in *Rhagoletis pomonella*. III. Male aggregation in response to an arrestant. Can. Ent. **104**: 275-283.

Rockwood, E. S. 1969. Enzyme variation in natural populations of *Drosophila mimica*. Studies in Genetics V, M. R. Wheeler, ed. Univ. Tex. Pub. **6918**: 111-132.

Smathers, G. A. 1968. A report on the microclimates in two Hawaiian kipukas. (In support of HAVO-N-8). Hawaii Volcanoes National Park. (Mimeograph).

Smathers, G. A. 1972. Plant succession, invasion and recovery on the 1959 Kilauea Iki eruption site, Hawaii Volcanoes National Park. Ph.D. Dissertation. Univ. of Hawaii, Honolulu.

Spieth, H. T. 1966. Courtship behavior of endemic Hawaiian *Drosophila*. In *Studies in Genetics III*, M. R. Wheeler, ed. Univ. Tex. Pub. **6615**: 243-313.

Stearns, A. T. 1966. Geology of the State of Hawaii. Pacific Books, Palo Alto, California. 266 pp.

Throckmorton, L. H. 1966. The relationships of the endemic Hawaiian Drosophilidae. In Studies in Genetics III, M. R. Wheeler, ed. Univ. Tex. Pub. **6615**: 335-396.

Wright, R. H. 1958. The olfactory guidance of flying insects. Can. Ent. **40**: 81-89.

Yoon, J. S., Kathleen Resch and Wheeler, M. R. 1972. Cytogenic relationships in Hawaiian species of *Drosophila*. II. The *Drosophila mimica* subgroup of the 'modified mouthparts' species group. In Studies in Genetics VII, M. R. Wheeler, ed. Univ. Tex. Pub. **7213**: 201-212.

Index